THE
REALM OF
MOLECULES

RAYMOND

DAUDEL

THE

REALM OF

MOLECULES

McGraw-Hill, Inc.

New York   St. Louis   San Francisco   Auckland   Bogotá
Caracas   Lisbon   London   Madrid   Mexico
Milan   Montreal   New Delhi   Paris
San Juan   São Paulo   Singapore
Sydney   Tokyo   Toronto

English Language Edition

Translated by Nicholas Hartmann
in collaboration with
The Language Service, Inc.
Poughkeepsie, New York

Typography by AB Typesetting
Poughkeepsie, New York

Library of Congress Cataloging-in-Publication Data

Daudel, Raymond.
    [*L'Empire des molécules*. English]
    The Realm of Molecules / Raymond Daudel.
       p. cm. — (The McGraw-Hill *HORIZONS OF SCIENCE* series)
    Translation of: *L'Empire des molécules*.
    ISBN 0-07-015642-5
    1. Atomic theory. 2. Molecular theory. 3. Biochemistry.
    I. Title. II. Series.
QD461.D2413 1993   92-24070
574.8'8—dc20

The original French language edition of this book
was published as *L'Empire des molécules*, copyright © 1991,
Hachette, Paris, France.
Questions de science series
Series editor, Dominique Lecourt

TABLE OF CONTENTS

INTRODUCTION

The idea that matter consists of atoms and molecules had to surmount almost unimaginable obstacles before being accepted by the scientific world. Today it is considered a long-established and obvious fact; but the legitimacy of the molecular realm—whose portrait Raymond Daudel paints with such masterful strokes—is not even a century old.

As recently as 1877, the illustrious chemist Marcelin Berthelot (1827–1907), who embodied for several generations those twin French ideals of scholarship and statesmanship, regarded "the atomic system [as an] ingenious and subtle fable," but one with no real justification. At the same time another eminent chemist, Henri Sainte-Claire Deville (1818–1881), was declaring that the idea of "molecules [which] merge together, repel and seek out one another" was "infantile." Adolphe Wurtz (1817–1884), a lonely partisan of the atomic theory, could find nowhere to conduct his research but at the department of medicine. In France, the "atomists" did not manage to penetrate into the teaching of science until the 1900s, and even in 1903, they felt the need to join together in a club (called "Molécule") in order to promote their ideas.

The results obtained since the beginning of the century nonetheless appear impressive. What has been called the "renaissance" of the atomic theory can be attributed to the English scientist John Dalton (1766–1844). This precocious genius, born the son of a weaver and simple agricultural laborer at Eaglesfield in Cumberland, became a professor at Manchester by the age of 26! Part of his fame is due to his research on a particular variety of color blindness called dyschromatopsia, from which he himself suffered; in Europe, the condition is now known as "daltonism." But science remembers him chiefly for the fact that in 1803, for the first time in the history of modern chemistry, he formulated the atomic hypothesis. The theory was taken up again in 1807 by Thomas Thomson (1773–1852) in a new edition of his *System of Chemistry*, and developed in the following year, once again by Dalton, in the first volume of his *New System of Chemical Philosophy*.

This hypothesis states that, because there is a limit beyond which an object cannot be broken down without causing it to lose its chemical identity, matter must be assumed to have a discontinuous structure, and one must assume that each chemical species has certain ultimate particles, or atoms. Dalton thus postulated that an "elemental" substance is made up of atoms that are all identical, and that the atoms of two different elements are also different. Hence two atoms of sulfur are identical, which is not the case for an atom of sulfur and an atom of gold. He concluded from this that a hundred or so different atoms must exist in Nature, and added that an atom of a given element could combine with one or more atoms of another element.

Contrary to a widespread belief—one bolstered, moreover, by a linear and positivist view of the history of scientific thought—Dalton did not construct this theory in an attempt to interpret the empirical laws of weight proportionality reported by earlier chemists. In fact, he approached the question through meteorology: his observations of auroras, trade winds, and the causes of rain led him to ask questions about the physical constitution of the atmosphere. Since he was not unaware of Lavoisier's work, and knew that air was composed of two gases with different weights, he made an audacious and purely speculative appeal to the atomic hypothesis of antiquity to explain why atomic weights had to be observed in the weight ratio of the components of a combination. He had the inspiration of taking the weight of hydrogen as a starting point (= 1), attempted to determine the atomic weight of oxygen, and suggested the use of symbols to represent atoms.

In 1811, the Italian Amedeo Avogadro (1776–1856) was the first chemist to use the word "mole-

cule" (meaning literally "little mass") to support a
hypothesis which stated that under identical temperature and pressure conditions, any gas would
contain the same number of particles within the
same volume. For many years there was no response
to this hypothesis, although André-Marie Ampère
(1775–1836) had immediately shown that it created
a consistent link between the results obtained by
Joseph Louis Gay-Lussac (1778–1850) on the
expansion of gases, and the essence of Dalton's theory. But it was not until the late 1850s that the
distinction between atoms and molecules—an
imprecise concept that had previously been a source
of numerous misunderstandings and uncertainties—
was finally established by Stanislao Cannizzaro
(1826–1910), then soon confirmed by Lothar Mayer
in his *Moderne Theorien der Chemie* [Modern theories of chemistry] in 1864. From this point on it
was understood that in representing chemical elements, the atom (the smallest part that can enter into
combinations) had to be distinguished from the molecule (the smallest part capable of an independent
and stable existence).

In the meantime, other research on the deter-
mination of atomic weights had added credibility to
the atomic theory. The Swedish chemist Jöns Jacob
Berzelius (1779–1848), a universally admired experimental chemist, was the pioneer in this field; he
published numerous tables including one (in 1826)

which listed values very close to those accepted today. In the same vein, work by two polytechnic students—Pierre-Louis Dulong (1785–1838) and Alexis-Thérèse Petit (1791–1820)—had established at the same time that all atoms have the same heat capacity, and that the atomic weight of a substance is inversely proportional to its specific heat, a finding first stated in 1819 and now known as the Dulong–Petit law.

The atomic weight table was confirmed by electrochemical measurements performed by Michael Faraday (1791–1867) in England. As an assistant to Humphry Davy at the Royal Institution, he had corresponded with Alessandro Volta, and made very good use of Volta's invention of the electric battery in 1800. In 1833, Faraday formulated the laws which now bear his name, and forged a link between electricity and chemical affinity.

Lastly, we cannot fail to mention the Russian Dmitri Mendeleev (1834–1907), who had worked at Wurtz' laboratory in Paris and proposed, in a communication to the Russian Chemical Society in 1869 entitled "Concerning the ratio between the properties and atomic weight of the elements," an initial periodic classification of the elements. His list of elements, arranged in the order of their atomic weight from the lightest to the heaviest, revealed that the same chemical behavior recurred periodically from one group to the next.

We already know the success achieved by Mendeleev, who did not hesitate to write: "No chemist had yet ventured to predict the properties of elements not yet discovered, nor to modify atomic weights or in general to regard the periodic law as a new law of Nature built on a solid foundation and capable of encompassing facts not yet generalized, as I have done from the beginning." In 1875, François Lecoq de Boisbaudran (1838–1912), utilizing the Russian chemist's table, discovered gallium, which he proudly named in honor of France; it was then followed by scandium (after Scandinavia) and germanium (after Germany). All of these elements had been "predicted" by Mendeleev. This already impressive edifice was completed in the 1860s by the beginnings of organic chemistry. Foreshadowed long ago by the successful synthesis of urea in 1829 by the German chemist Friedrich Wöhler (1800–1882), this discipline counted among its founders the Russian Alexander Boutlerov (1829–1896), Archibald Couper (1831–1892) in Scotland, and the German scientist August Kekulé (1829–1896), all contemporaries of Berthelot. By a variety of routes, their work established the general principle that organic molecules consist of a chain of carbon atoms attached to one another by four bonds.

Given this situation, opposition to the atomic theory by Marcelin Berthelot and his acolytes began to look like a rear-guard action, inspired by an "official" philosophy (positivism) which stated that science must not make hypotheses about anything it cannot see. And in fact, no one had ever seen an atom, or a molecule. The support of French political power transformed this battle into a scientific and industrial disaster. Jean Jacques has remarked on several occasions that while Berthelot was hurling his anathemas, the atomic theory was becoming an effective instrument in transforming the industrial production of dyes, perfumes, and drugs, especially in Germany and England.

But this inglorious episode, replete with lessons for the comparative sociology of science, would not be worth a second glance if it did not illuminate a difficulty that has been astonishingly neglected. The atomic theory did not become finally established until the physicists took up the theme first proposed by chemists. Here again, the story is just as tortuous, because until the end of the 19th century, physics was in no shape to provide any support. Since 1850, reeling from the formulation of the Second Law of Thermodynamics, it had been undergoing a "crisis" which some felt was calling into question the very value of modern science.

This law was stated by Sadi Carnot (1796–1832), son of Lazare Carnot, in 1824 in his *Réflexions sur la puissance motrice du feu et sur les machines propres à développer cette puissance*

[Reflections on the Motive Power of Fire and on Machines Fitted to Develop That Power], but its immense scientific implications were not discovered until 1851 by William Thomson (1824–1907)—later raised to the peerage with the title of Lord Kelvin— who in turn had been alerted by a paper published in the *Journal de l'École polytechnique* by Émile Clapeyron (1799–1864).

The German physician Julius Robert von Mayer (1814–1878) and the English industrialist James Prescott Joule (1818–1889) had established during the 1840s (quite independently of each other), the "first principle," that of the equivalence of heat and work. Carnot, for his part, followed by the German physicist Rudolf Clausius (1822–1888), had revealed the irreversibility of processes by which heat is converted into work. But this irreversibility, which Clausius called "entropy," seemed to contradict the fundamental tenets of classical mechanics, in which all phenomena were assumed to be reversible!

From this ensued the crisis, which led some very eminent scholars to reject the atomic concept or at least to assign it only symbolic value, since they associated the idea of the atom with the mechanistic notion of matter and motion that had prevailed since Newton. For example, the great physicist Ernst Mach (1838–1916) expressed very strong reservations about the atomic theory in his famous *Mechanics*: it was, he explained, simply a "mathematical model for describing the facts," a provisional model that should be replaced as soon as "a more natural concept" came along. The German chemist Wilhelm Ostwald (1853–1932) provoked a huge uproar in the *Revue générale des sciences* in 1895 when he published his article entitled (by the editors) "The defeat of contemporary atomism." In a sarcastic vein, he took issue with anyone who "repeats, as an axiom, that only the mechanics of atoms can provide the key to the physical world"; he railed against the supposition that "the highest plane that our explanation of the world can attain is to reduce it to a system of moving material points." The crisis was not resolved until 1905, when Albert Einstein (1879–1955) scrubbed classical mechanics clean of its philosophical presuppositions. With the work of Jean Perrin (1870–1942), the physical reality of atoms was no longer in doubt, although it proved to be very different from the "moving material point" so justly ridiculed by Mach and Ostwald.

By confirming the thesis of the atomist chemists, physics in a sense was simply reclaiming its own due, because in fact, long before its controversial rebirth in chemistry, the atomist hypothesis had been advanced many years ago by physicists. Let us now take a reasoned and measured look at a history which, although it extends over millennia, has nevertheless remained essentially hidden.

Conventional wisdom assigns credit for the very first of the "atomist" concepts to two pre-Socratic thinkers, Leucippus of Miletus and Democritus of Abdera, who lived in the 5th century B.C. It is certain that Leucippus, in order to explain the reality of change—the possibility of which had been denied a century or so earlier by Parmenides in his famous poem on Being—became what Aristotle called a "physiologist": he investigated Nature. (The terms "philosopher" or "physicist" cannot be applied to him without anachronism, any more than they can to the other thinkers of his era.) His investigation stemmed from the principle that "nothing happens by chance, but everything happens for a reason and by necessity." This was an unquestionably anti-religious stance, as clearly demonstrated by G. E. R. Lloyd, who recalls the accusations of "atheism" that Plato, a century later, would try to pin on the physiologists.

Leucippus developed a doctrine stating that nothing was real in Nature except atoms and the void. All differences among physical objects, whether qualitative or quantitative, were the result of changes in the shape, arrangement, and position of atoms. Each individual atom was thought to be unengendered, unalterable, homogeneous, undeformable, and, as the Greek word itself implies, "indivisible" (α = not, $\tau o \mu o \varsigma$ = cut or slice). According to Leucippus, an infinite number of atoms can be imagined, and between them must necessarily exist the void, without which they could not move, collide, rebound, or arrange themselves to form, temporarily, some substance.

This system was extended by Democritus, who had dared—in the face of the most obvious evidence and the most sacred certainty—to deny the divine nature of the stars and believe, for example, that the Sun was "incandescent iron or a flaming stone." He enriched this framework with a doctrine of sensations which, according to Theophrastus, attributed the perceptible qualities of substance to the shape of their constituent atoms: thus angular atoms produced an acid taste, round atoms a sweet taste, etc.

Whether we want to or not, it would be impossible to draw a straight line from this grandiose system to modern atomism, either Dalton's theory or the ideas that have become established since the beginning of this century with the "quantum revolution." The ancient concept of the atom does not correspond to the idea of an element adopted by 19th-century chemistry: the idea of a finite number of atoms, involving particular properties, is radically foreign to it; and moreover the Greek atom is by definition indivisible and homogeneous, whereas the contemporary atom, far from presenting itself as a "little body," emerges as a highly complex reality whose components were in fact revealed only by "dividing" it into smaller and smaller fragments with more and more powerful tools.

It might be objected that Dalton, like contemporary physicists, insisted on acknowledging Leucippus and Democritus as his distant ancestors. To understand the reasons behind this reference, it is not entirely profitless—however paradoxical it might seem—to remember the unhappy fate that befell atomist philosophy. Its anti-religious tinge, already mentioned, exposed it to the virulent criticisms of Plato. Aristotle, although more moderate in tone, ultimately proved no less implacable. These two princes of Greek philosophy preferred to follow another tradition with regard to the identity of the constituent "elements" of Nature, one that began (as far as we know) with Empedocles of Agrigentum, a contemporary of Leucippus. Once again taking issue with Parmenides on the problem of change, Empedocles proposed to rely on the testimony of his senses to identify what he called the "roots" of all natural beings: earth, water, air, and fire.

This theory of the four elements, believed to correspond to the opposing perceptible qualities of dry, wet, cold, and hot, was taken up by Plato in the *Timaeus*, but geometricized in accordance with a Pythagorean concept in which each of the four basic substances was identified with one of the regular solids. Aristotle subscribed to this line of anti-atomist thought when he, too, declared that earth, water, air, and fire represented the constituent matter of everything that existed on Earth. The four "elemental" substances were, in his eyes, linked to two pairs of primitive opposites: hot and cold, and wet and dry. He rejected Plato's geometrism, and introduced a fifth element, the ether, to which he ascribed the circular and regular motion of the stars. It is very obvious from this that, despite his constant protestations of empiricism, the essence of Aristotle's "physics"—an integral part of his philosophy alongside ethics and logic—is in fact subordinated to ideas of a metaphysical and theological nature.

Although it was criticized and corrected in matters of detail by Theophrastus (371–332 B.C.), Aristotle's successor as head of the Lyceum, and by others after him, the theory of the four elements was passed down through the millennia. We find it in the *Natural History* of Pliny, who perished in the eruption of Vesuvius in 79 A.D., and who created a vision of Nature, rocks, plants, and animals that would be shared by fifteen centuries of human beings—scholars, artists, priests, and philosophers—and we find it especially in Stoic philosophy, where fire played a very active role. These ideas would serve as "scientific" support for the medicine of Claudius Galen in the 2nd century, and endure through the Middle Ages to the 17th century, having survived a flare-up of mysticism during the Renaissance.

Beyond any doubt, this extraordinary hold on the human spirit resulted not from any supposed "realism," but from the oneiric and poetical value embodied in the concept of an "element." Gaston Bachelard looked into the works of the alchemists and the locutions of the inspired poets, and masterfully deciphered the symbols of this fantasy, which emerges in his words as the very root of articulate humanity.

The theory also owed its success to the exceptional philosophical and social destiny of Aristotle's thought. The power of his encyclopedic output—providing, for every question, an answer that was consistent with all of his doctrine—and the immensity of his empirical inventory, in natural history as in geology or astronomy, explain its seminal and perennial dominance in scholarly thought. It is also unquestionably true, as emphasized by Robert Lenoble, the great historian of science, that Aristotle's theory was the first one capable of rationally dismissing any magical vision of Nature. But an essential part of its good fortune has been its ability, ever since the 13th century and the immense contribution of St. Thomas Aquinas (1225–1274), to adapt to the dogmas of Christian theology—provided, of course, it was modified on certain fundamental points. When Plato was referred to as counterpoint, it was only in connection with the "demiurge" that was expressly addressed in the *Timaeus*.

In any event, the theory of the four elements prevailed against the atomist thesis—which, with the impact imparted to it by Epicurus (341–270 B.C.) and the somber grandeur that it achieved in the stanzas of Lucretius' *De rerum natura* (95–55 B.C.), was considered (with good reason) the very essence of a subversive philosophy because it was so resolutely anti-religious. And hence it always remained the doctrine of a few, and the butt of slander.

Each of the decisive moments at which modern science had to detach itself from Aristotelianism in order to create its own concepts has therefore been marked by an appeal, explicit or otherwise, to the atomist tradition—despite the fact that, from Epicurus to Lucretius, Galileo, and Newton, it has never had a single "fact" to sink its teeth into. It was not that the reality of Democritus' atom seemed a steadier base for the physical sciences, but rather that an affirmation of the shared dominion "of chance and necessity" opened up in the face of teleology, anthropocentrism, and creationism, new avenues for that "investigation of Nature" that physics in fact became.

Of course the history of science cannot be simplified as representing two lines of parallel philosophical thought that developed by maintaining a constant polemic between them. Many thinkers tried to find a compromise, among them René Descartes (1596–1650), who could not stretch his

Epicureanism so far as to accept the idea of the void, and recoiled from the concept of an infinite Universe. Robert Boyle (1627–1691), son of the first Count of Cork in Ireland and one of the founders of the Royal Society, proved more radical when, in his famous book entitled *The Sceptical Chymist* (1661), for the first time he detached the "element" from its Aristotelian heritage. Harking back to the atomist tradition, he defined this concept as the elementary quality of any substance that could not be decomposed by experimental means, thus opening the way for the founders of modern chemistry. But then Newton in a sense re-established the Cartesian compromise, seeking in God, "Lord of the Universe," the metaphysical underpinning for immediate action at a distance, which the laws of his particulate Universe could not in themselves explain.

What has come down from the atomist tradition, therefore, is by no means a concept that remained unchanged, nor a specific notion of matter; rather it is a philosophical position which, in defiance of Aristotle, denies the preexistence and primacy of form—and thus of sensation—over matter; and in the same philosophical gesture affirms the necessity of natural processes and treats as merely contingent the fact that they are governed by laws that we formulate. That contingency can, moreover, be imputed to an act of the Creator.

Guided by one of the masters of contemporary chemistry, you will now discover the immense territories of knowledge conquered by this philosophical approach, now infused into more and more powerful experimental devices, in many areas including the life sciences. You will ultimately see how what Raymond Daudel calls the "quantum molecular sciences" are renewing and invigorating the human spirit. Air, water, earth, and fire will surely always sustain the imagination, but the undulating molecule, in its dance of a strange and spellbinding modernity, has now contributed its own vibrant flame.

Dominique LECOURT

I

FROM THE INANIMATE

TO THE LIVING

AT THE HEART OF INANIMATE MATTER

The concepts of "atom" and "molecule" may appear abstract, quite remote from our daily lives; nevertheless they play a central role in our understanding of the world and our ability to control our environment. We know today that they represent the constituent elements of both inanimate objects and the living beings with which we are familiar. We have discovered that we ourselves are made of atoms and molecules.

If we agree, for simplicity's sake, to define an atom as a monatomic molecule—in other words, a molecule reduced to a single atom—the molecule becomes the fundamental element of all the objects that create the context for our life on Earth. The soul of inanimate objects, so dear to the French poet Lamartine's heart, arises from the life of molecules. Because in a sense, they are alive: even the molecules of inanimate objects move, meet, attract and repel each other, change, and die, giving birth to new molecules. It therefore seems no exaggeration to say that we live in and under the realm of the molecules.

Gases consist of molecules moving in the "immense void," to use Lucretius' lovely expression. The oxygen in air, without which life would be impossible for us humans, is in turn composed of diatomic molecules, that is, molecules composed of two oxygen atoms (O_2). The noble gases such as neon and argon are made up of free atoms, in other words monatomic molecules. The air we breathe is a very complex mixture whose principal constituents are oxygen, nitrogen, the noble gases, water vapor, and carbon dioxide, which in turn consists of molecules that each contain one carbon atom bonded to two oxygen atoms.

At atmospheric pressure and room temperature, the number of molecules contained in one cubic centimeter of gas is approximately 30 billion billion. These molecules are constantly moving, traveling at speeds that can reach several hundred meters per second. Since they are very numerous, they are always colliding. Just imagine that a molecule bumps into its neighbors about 10 billion times a second, and you will understand that its trajectory is extremely complex. All these movements together constitute thermal agitation, which becomes increasingly frenetic as the temperature rises.

Now let us consider a certain quantity of gas, or a certain number of molecules, enclosed in a cylinder in which a piston can slide. Pressing the piston into the cylinder reduces the volume of the gas, and thus forces the molecules to hit each other even more often, and to press against one another. We say then that the "pressure" of the gas has increased. If the temperature is not too high and if we continue to raise the pressure, a time will come when the molecules touch one another. They begin to roll against each other (a little like the elements in a ball bearing, but with more freedom). The gas has become a liquid; that is why, when the temperature is low and the air is full of water vapor, fog blurs the outlines of everything around us.

A liquid is thus a collection of disordered molecules, like those in a gas, but pressed against one another like "little marbles piled up at the bottom of a bag," to use André Guinier's colorful expression. Exactly as they do in a gas, these molecules move and change position as they rotate, slide, and roll. Every second, a water molecule at 20°C (68°F) will move an average of 1/25th of a millimeter. At 50°C (122°F), the same molecule will have covered twice that distance. In a gas such as air, at a temperature of 20°C, the molecules cover distances approximately 150 times as great in the same time period. Conversely, when a liquid's temperature is lowered, its thermal agitation decreases. Below a certain temperature, the molecules essentially no longer change their positions; they simply vibrate and rotate around a fixed point. The liquid has become what we call a solid. In the case of water (again at atmospheric pressure), this solidification produces ice at the familiar temperature of 0°C (32°F).

When a piece of sulfur burns in air, its molecules change and die. Combining with oxygen molecules in the air, they create new molecules of SO_2 containing one sulfur atom and two oxygen atoms and form a gas with a suffocating odor called sulfur dioxide. This gas is believed to be responsible for the acid rain which is devastating our forests, since it combines with the water in clouds to form sulfuric acid. All the molecular transformations in which we find some molecules dying and others being born are called chemical reactions. But we must not fix our gaze exclusively on pollution, as so often happens today, and regard chemistry as something malevolent or harmful; because it is by way of reactions like these that we humans manufacture most of the products, materials, and medicinal agents that we need. Our very lives are the result of billions of chemical reactions taking place in our bodies every second.

AT THE HEART OF LIVING MATTER

Every living being—every animal, plant, bacterium, and virus—is made of molecules, some of which are identical to those in non-living matter. There is no difference, for example, between the water molecules in our blood and those in a mountain stream. But living matter contains specific molecules that are not normally found in inanimate objects (except in fossils). These particular molecules of life, which are often large and complex, can contain thousands or even millions of atoms. So understanding life becomes mostly a matter of solving a very difficult chemistry problem.

To develop, grow, and survive, any living being must extract from its environment the molecules it needs. And to manufacture from these molecules (or "synthesize" from them, as the chemists say) the ones necessary for life, the living being needs energy. A plant will find that energy in the abundant light radiated from the Sun onto the Earth, while a mammal will obtain energy by "burning" its food. It is estimated that every human being continuously emits a quantity of heat equivalent to the output of a 200-watt electric radiator, while maintaining a body temperature of about 37°C (98.6°F).

The green leaves of plants contain tiny organs called chloroplasts that are rich in large molecules of a substance known as chlorophyll. These molecules can be described as tiny antennas spread out towards every slight ray of light, in order to capture its energy. Other molecules fix carbon dioxide from the air, and thanks to the energy extracted from light, a chemical reaction occurs involving carbon dioxide, water, and the phosphates in the plant's sap. That reaction allows the plant to synthesize certain important molecules that each contain one atom of phosphorus, three carbon atoms, four oxygen atoms, and six hydrogen atoms. This molecule is the starting point for the synthesis of other, more complex ones that the plant needs in order to live and grow.

Green plants therefore feed by extracting matter and energy from the soil, the air, and solar radiation. Herbivores eat the plants, and carnivores eat the herbivores. Humans eat both plants and animals. As Jacques Ruffié has written: "A tree will never have an opportunity to eat cold cuts, but while some of our contemporaries, stretched out nude (or nearly so) on every Mediterranean beach in August, may hope to get a tan by manufacturing melanin [very dark-brown molecules], none of them will ever manage to synthesize chlorophyll." In other words, the distinction between plants and animals lies mostly in the way in which they take advantage of their environment, that is in the nature of the chemical reactions they can perform.

Bacteria behave more ambiguously. Some of them feed like plants, but the ones we know best— the tuberculosis and leprosy bacilli, the staphylococci that cause boils, the gonococci responsible for venereal diseases—essentially live as parasites in animals like us. The viruses, in turn, parasitize anything that lives, including the bacteria from which they probably derived. Examples include the tobacco mosaic, which attacks the leaves of the tobacco plant; and the chicken-pox virus, which can often hide for long periods in sensitive ganglia, emerging sometimes thirty, forty, or fifty years later to produce shingles. The AIDS virus destroys our immune defenses, leaving the body helpless against its usual aggressors such as helminths, bacteria, other viruses, protozoa, fungi, and cancer cells. Numerous "opportunistic" infections then ensue, one or more of which causes death. All of these conditions involve chemical reactions. In fact, just as we live thanks to chemical reactions, most often we die because of other chemical reactions.

THE CELL: A CHEMICAL FACTORY

Have you ever slid a scalpel under the skin covering a slice of onion, detached a little scrap, and looked at the cells through a simple microscope? If you have, you probably remember seeing more or less regular hexagonal shapes with a little rounded object near the center. Those hexagonal elements are called "cells," and the dot at their center is the "nucleus" (not to be confused, obviously, with the nucleus of an atom, which is infinitely smaller).

The cell represents the fundamental unit of living matter. Every living being, with the exception of viruses, is made of cells. The simplest such beings consist of only a single cell, like bacteria and protozoa. But most living beings contain a very large number of cells; one gram of human tissue, for example, contains about a billion cells!

The dimensions, shapes, and structure of cells are extremely varied, since they play very different roles in an organism. A bone cell, a liver cell, and a brain cell perform completely different functions. The most common cells are between 10 and 100 thousandths of a millimeter in diameter (0.01–0.1 mm). But there are some enormous ones, like birds' eggs, each of which consists of a single cell. Liquid tissues such as blood also contain many cells of varying kinds: red cells, white cells (especially macrophages, lymphocytes, and polynucleates), and platelets. One drop of blood may therefore contain millions of red cells and thousands of white cells.

Most cells (but not all) multiply and divide during their lifetimes, either as they grow, or in order to replace dead cells. In François Jacob's light-hearted phrase, "The dream of every cell is to become two cells." When the single cell of a bacterium divides, it yields two cells identical to the first one. This is the basis of the phenomenon of bacterial reproduction. In humans, the initial ovum splits into two cells, then each of these into two others (not necessarily identical), thus forming the embryo, fetus, newborn, child, and finally the adult. In this case, cell division constitutes the major process by which the living being develops.

The chemical reactions that permit these cell divisions are complex. Let us take the very simple case of the division of a typical cell like the one in *Ascaris*, a little worm that lives as a parasite in the human intestine. As the process begins, we see little rods—located in the nucleus at the center of the cell, and called chromosomes—getting thicker. Then two spindle-shaped objects, each starting from a tiny ball, enclose the chromosomes. Later we see the chromosomes divide into two distinct packets. A membrane forms between the two packets, and the initial cell produces two new cells. This phenomenon of "mitosis" can be observed with an ordinary microscope. At the molecular level, however, it corresponds to a very large number of highly complex chemical transformations.

Two types of molecules play an especially important part in these chemical reactions which underlie all life: nucleic acids and proteins. The name "nucleic acid" was given to the first category because they are found in large numbers in cell nuclei. One of these acids—deoxyribonucleic acid, or DNA—turns out to be absolutely fundamental. It carries the genetic program, the agent of heredity and the essential characteristics of the species, be it a bacterium, a plant, or an animal. In each cell, this program contains the list of chemical reactions that the cell must perform in order to contribute to the life of the organism of which it is a part.

The architecture of this molecular edifice is breathtaking. DNA essentially consists of two strands located side by side, joined together in the form of a helix. These two molecular strands contain atoms of carbon, oxygen, hydrogen, and phosphorus. They are linked by groups of atoms called bases, which can exist independently as little molecules. DNA contains four different bases: adenine, thymine, cytosine, and guanine. These are conventionally symbolized as A, T, C, and G, respectively. The two DNA strands are linked together by two—and only two—types of base pairs: either an A–T pair (adenine–thymine) or a C–G pair (cytosine–guanine). In other words, an adenine can only be paired with a thymine, and a cytosine only with a guanine.

When a cell divides, the chromosomes thicken and then divide in two: DNA acts as the orchestra conductor in this process. When cell division begins, the two strands of DNA move apart; each adenine then pairs with a thymine, each thymine with an adenine, each cytosine with a guanine, and each guanine with a cytosine. Consider, for example, a fragment of DNA whose two strands contain the following message:

$$-G-C-A-A-$$
$$-C-G-T-T-$$

When the two strands move apart, this fragment will produce two unattached messages: –G–C–A–A– and –C–G–T–T–. These messages, as they encounter free bases in the cellular medium, will then attract their respective complements to form exclusively A–T and C–G pairs. The free –G–C–A–A– fragment will become:

$$-G-C-A-A-$$
$$-C-G-T-T-$$

and the –C–G–T–T– fragment will become:

$$-C-G-T-T-$$
$$-G-C-A-A-$$

So the initial fragment:

$$-G-C-A-A-$$
$$-C-G-T-T-$$

will have produced two fragments identical to itself. It will have duplicated itself, which is what causes the chromosomes to thicken. One cell will then be able to give birth to two daughter cells identical to the parent cell.

To make a little more progress towards understanding this very complex event, we need to bring in a second category of molecules specific to living things: proteins. Some of them, called enzymes, catalyze (that is, facilitate or make possible) certain chemical reactions that are essential for cellular life. Others constitute hormones, molecular messengers that move through the organism to give orders to certain organs. For example, in human beings the pancreas manufactures a protein hormone called insulin which regulates the concentration of glucose in the blood. It is used to treat certain forms of diabetes. Another type of protein is hemoglobin, which is abundant in our red blood cells; we know that it transports oxygen from the lungs to our cells, where this gas is used to "burn" our food as part of the series of chemical reactions referred to as the respiratory chain. This series of reactions provides the body with some of the energy it needs.

The membranes of our cells contain a number of proteins with a wide variety of functions. François Chapeville believed that proteins were the most important constituents of living organisms: "They imprint upon the cell, indeed upon the entire organism, its form and its function." We will therefore first summarize what they are and then discuss how they are synthesized or manufactured. Proteins are made up of a series of some twenty distinct molecules called amino acids, specifically: glycine, alanine, valine, leucine, isoleucine, phenylalanine, tryptophan, histidine, proline, threonine, tyrosine, hydroxyproline, lysine, arginine, aspartic acid, glutamic acid, cysteine, and methionine. Our body obtains these raw materials by digesting foods that contain proteins manufactured by plants or other animals. Each protein contains a well-defined number of amino acids attached to one another (or, as chemists say, combined or "polymerized") in a strictly determined order. A given protein does not necessarily include all 20 amino acids. Without giving a complete description of the marvelous mechanism which allows cells to manufacture proteins, I can at least emphasize certain of its characteristic features. The recipe that must be followed in order to trigger the chemical reactions necessary for production of a particular protein is embedded in a message that resides in a fragment of DNA called a "gene." The set of all genes constitutes the "genome" of the particular living being in question.

The cell's way of reading the recipe contained in a gene is to decode the message by starting at the beginning of the gene and locating each group of three bases. The combination AGC, for example, appears to be an order to use a tyrosine; while AAT is the instruction for an alanine. If the cell "reads" –A–G–C–A–A–T–, it will attach an alanine to a tyrosine, and so on. Reading the gene thus activates that series of chemical reactions which hooks together, in the correct order, all the amino acids (several hundred, or even several thousand) needed to construct the architecture of a particular protein such as insulin.

Each individual has a particular genome, and will therefore manufacture particular proteins which conform to the "recipes" in his or her genome. That is why one man may be tall and another short, one woman dark-skinned with brown eyes and another fair-skinned with blue eyes.

THE HEREDITY LOTTERY

Like all living things, we are essentially molecular beings: every act in our lives is accompanied by chemical reactions, beginning with the ones by which our species reproduces. The first step should therefore be to identify the structure of the genome found in each cell of a man or woman. Because of the exceptional importance of this structure, an international group of research teams (Human Genome Project) is currently working (with several billion dollars in funding) to determine every detail of it. Each human cell contains 23 pairs of chromosomes. In women, the two chromosomes of each pair contain genes which are similar but not identical; the twenty-third pair consists of two chromosomes called "X" chromosomes. This is called an XX pair. In men, the chromosomes of a pair are also similar, except in the case of the twenty-third pair, which contains one X chromosome (just as in women), but also a much smaller chromosome called a Y.

In short, the twenty-third pair of chromosomes is XY in men and XX in women. The sexual differences between men and women are based on this fundamental distinction in the structure of this twenty-third chromosome pair. As a result, a woman has two ovaries, two Fallopian tubes, a uterus, and a vagina; and a man has a penis and two testicles.

By means of a very specific process called "meiosis," the sexual organs produce original cells—called gametes—which contain only 23 chromosomes, rather than 23 pairs. The genome in a gamete thus comprises half the genes of the individual in question. In women, one of the follicles in the ovary manufactures a gamete called an oocyte. Released into the Fallopian tube in the middle of the estral cycle, this gamete slowly moves down the tube, then falls into the uterus. It of course contains one X chromosome, produced by splitting the XX pair. In men, the testicles manufacture gametes called spermatozoa, but since a man's cells contain an XY twenty-third pair, the gametes will have either one X chromosome or one Y chromosome. Approximately half the spermatozoa will be X, the other half Y.

Sexual desire in women is primarily controlled by a hormone manufactured by the ovary, called estrone; men's libido is governed by a hormone manufactured by the testicles, called testosterone. When a man and a woman unite in the act of sexual intercourse, at the moment of his climax the man releases several million spermatozoa into the woman's vagina; they eventually reach the cervix at the neck of the uterus, swim up into the uterus and enter the Fallopian tubes. If one of these spermatozoa encounters, in one of these tubes, an oocyte just released by the woman, fusion between the oocyte (which contains 23 chromosomes) and the spermatozoon (which also contains 23 chromosomes) will form a fertilized cell called the zygote, which will comprise 23 *pairs* of chromosomes; half the genes will come from the father, and the other half from the mother. If the spermatozoon contains an X gene, the fertilized egg will contain an XX pair and will develop into a girl. If the spermatozoon contains a Y gene, the egg will be XY and will produce a boy. That is why, in our species, males and females are born in approximately equal proportions.

A portion of a baby's future is therefore already written in the fertilized egg at the moment of conception. We can indeed refer in this context to a "heredity lottery," since during the division process which leads to the production of gametes, the genes of a single pair of chromosomes mix at random in the generating cell, a bit like shuffling a deck of cards before cutting it into two equal halves. To put it another way, every spermatozoon does indeed contain half of a man's genes, but two different spermatozoa produced by the same man do not contain the same half. Similarly, each oocyte released by a woman contains half her genes, but two different oocytes released by the same woman do not contain the same genes. Every fertilized egg always gets half its genes from the father and half from the mother.

But two different eggs produced by the same couple comprise different halves of genes from the mother and the father. That explains why each man and woman possesses a separate physical personality, distinct from that of everyone else—with the single exception of true or identical twins. Identical twins are born from the same egg, which just happens to divide into two eggs at the beginning of gestation. Identical twins therefore contain the same genes. That is why they are very much alike physically, and even often psychologically, although to various degrees; only differences in their environment and social history can create significant discrepancies.

Thus, after the random events that accompanied the mixing of chromosomes when the gametes were formed, much of the future of the child and therefore of the adult is predetermined in the egg as soon as conception occurs. In a way, the fertilized egg already possesses a memory of its future. This fact is so true, and so important, that the mere presence of a certain gene can indicate a predisposition for a certain specific illness. There are hundreds of diseases whose origins are inscribed in the genome. Millions of human beings suffer from these genetic diseases. A new branch of medicine is now rapidly developing: "predictive" medicine, which is beginning to allow doctors to predict these illnesses while the fetus is still in the womb; the hope is that soon it will be possible to prevent them, or at least to mitigate their most unpleasant consequences.

OUR MOLECULAR SELVES

In Molière's satirical play *Le Bourgeois Gentilhomme*, the pompous Monsieur Jourdain finds that all his life he has been speaking in prose without realizing it. Similarly, human beings, since their conception, have been "doing" chemistry without realizing it. From the first moments of their intrauterine life until they die, and even afterwards, men and women unconsciously function as extremely complex, highly sophisticated, and very powerful chemical factories.

Digestion of our food, for example, brings into play a huge number of chemical reactions. As soon as we eat a piece of bread, we perceive a sugary taste as the enzymes in our saliva convert starch into sugar. Then gastric pepsin, pancreatic lipase, and intestinal juices continue the chemical transformation of our food into molecules that can be directly used by our body, that is by our cellular machinery.

The respiratory cycle allows us to "burn" some of these molecules formed from our food, and to produce the energy needed to operate these cellular factories. Our cells also contain tiny organisms called "mitochondria," which can store this energy for subsequent use. The operation of our brain, even the most fleeting of our thoughts, are accompanied by innumerable chemical reactions. The impulses circulating through our nervous systems are a sort of electrical current that leaps from one nerve cell to another through the intermediary of neurotransmitter synapses. The vivid imagery of Jean-Pierre Changeux' description is quite appropriate: "The encephalon (the essential part of our nervous system) appears before us as a gigantic assemblage of billions of neuronal spider-webs entangled in one another, through which ripple and propagate myriads of electrical pulses echoed here and there by a rich palette of chemical signals."

When our sense organs function, a multitude of chemical reactions occur. Our retina cannot see light without photochemical reactions. When we give a command to a muscle, chemical reactions carry it out. When an intruder (bacteria, virus, fungus, etc.) enters our body, an army of cells goes to war against it, thanks to a whole series of chemical phenomena. Human beings are therefore indeed, above all, molecular beings whose very lives depend on chemical reactions, although we must not be so eager or so complacent as to reduce them to simple machinery. A profound understanding of the world around us, whether inanimate objects or living beings, turns out to be possible only to the extent that we understand molecules.

The quantum molecular sciences, which we will now discuss, provide the only key that will allow us to enter and control this fascinating field of knowledge, and to use the powers that it gives us.

II

QUANTA:

FROM ATOMS

TO MOLECULES

*WAVE–PARTICLE DUALITY
AND ENERGY QUANTA*

If we want to study the motion of a train or an automobile, or to calculate the trajectory of a rocket that will take astronauts to the Moon, we have an entire mathematical arsenal at our disposal: classical mechanics, created by Isaac Newton and developed during the 18th and 19th centuries. At the end of the 19th century and the beginning of the 20th, however, certain phenomena were discovered that could not be understood on the basis of classical theory. Scientists had to admit, not without difficulty or apprehension, that classical mechanics could not account either for the properties of atoms and molecules, or for the phenomena which result from interaction between matter and light.

This genuine impasse faced by classical physics led scholars to create an entirely new science that represented a true leap into the unknown, a "revolution" that led to the establishment of the quantum theories. The history of the theory of light offers an easy and historically legitimate introduction to the quantum theories. Let us remember that, for Lucretius, light consisted of little particles, or atoms. In the 17th century, Descartes picked up this idea and postulated that light propagated in straight lines, called light rays. With this hypothesis, he was able to formulate the laws of reflection and refraction.

A little later, Newton proposed the first mechanistic theory of light by supposing that every light source hurls into space a multitude of very fast, very light little particles; because of their nature, they follow rectilinear paths. When they enter our eyes, these little particles produce sensations of light. Newton's concepts thus created a sort of synthesis between the ideas of Lucretius and Descartes, and provided an explanation for every phenomenon associated with light—until the day that a new phenomenon was discovered. That phenomenon was interference, illustrated by the curious fact that dark areas can be produced on a screen by illuminating it with two different light sources. The rays become superimposed (we say that they "interfere" with one another); in other words, in certain cases, light + light = darkness.

It is very difficult to understand the reason for this paradox by simply applying Newton's theory. Augustin Fresnel (1788–1827) was therefore led to think of light as a wave phenomenon. He imagined that a wave, similar to the one that travels along the surface of a body of water when a stone is dropped into it, propagates along the entire length of a light ray. The "frequency" of this wave is the number of oscillations that it makes every second, and its "wavelength" is the distance separating two adjacent crests (or troughs).

It is easy to see that "interference" can be produced on the water's surface by dropping two stones at the same time. One wave spreads out from each point of impact. At certain places the water's surface does not move, since one of the waves that would have caused a trough cancels out the movement due to the other, which would have created a crest. At other places the opposite occurs: when two troughs meet, they form a deeper trough; when two crests are added to one another, a higher wave appears.

With Fresnel's wave hypothesis, it became possible to explain every light-related phenomenon, including reflection and refraction. The particle theory then lapsed into oblivion. On this same foundation, scientists even managed to construct an initial theory of color, demonstrating that the color of light radiation depends on its frequency. In particular, it was noted that as one moves from red to violet through the colors of the rainbow (red, orange, yellow, green, blue, indigo, and violet), the frequency of the corresponding light doubles. The frequency of red is on the order of 400,000 billion oscillations per second, while that of violet is about 800,000 billion.

Towards the end of the last century, the discovery of the photoelectric effect once again called everything into question. Someone noticed that light could strip electrons away from an atom of metal. It was therefore possible to set up an experiment so as to produce an electric current with photons. But the curious thing was that for a given metal, the phenomenon only occurred if the frequency of the light being used was above a certain value, a certain threshold. Below that frequency, the metal could be flooded with light and nothing would happen.

To integrate this strange observation into existing physical theory, Albert Einstein turned to a hypothesis advanced several years earlier by Max Planck to explain another radiation phenomenon, and thereby forged a synthesis between the ideas of Newton and Fresnel. He assumed that light was made up of particles (called photons), and that their propagation was proportional to the frequency f of the wave, according to the formula $E = h \times f$ (where h was the constant introduced into physics by Planck). To interpret the threshold of the photoelectric effect, Einstein supposed that only a single photon was involved in knocking an electron off an atom. If the energy of that photon was less than the energy needed to remove the electron, nothing would happen no matter how many photons were striking the metal. That would be the explanation for the threshold.

Thus, led by Planck and Einstein, physicists were forced to recognize that the energy of a beam of light is concentrated in particles that each contain a "quantum" of energy, in other words a tiny quantity of energy proportional to the frequency of the light wave. Of course this theory was applied to every radiation with the same characteristics as light: radio waves, infrared, ultraviolet, X-rays, and gamma rays. Scientists found that the phenomenon of light had a dual interpretation, both wavelike and particlelike. Each ray of light became simultaneously the trajectory of a photon *and* the pathway of a wave.

In 1923, Louis de Broglie suggested that this duality could be generalized to every particle. He advanced a hypothesis stating that the movement of any particle (electron, atomic nucleus, photon, etc.) was accompanied by the propagation of a wave which guided it. He wrote in his thesis: "This idea that the movement of a material point always conceals the propagation of a wave needs to be studied and perfected; but if it could be given an entirely satisfactory form, it would represent a synthesis of great rational beauty."

But when these theories were applied to the interpretation of one particular light-related phenomenon—the diffraction effect—some radically new ideas emerged. When a beam of light strikes a very small hole, it "diffracts": for all intents and purposes, the hole becomes a light source, since light rays emerge from it in all directions, not just in the direction of the initial beam. What is odd is that the diffraction effect occurs even if the light intensity is so low that the photons pass through the hole in single file, one by one.

How can the trajectory of one photon be correlated with the paths of a wave, since at a given moment there is only one photon trajectory and a whole series of light waves? Theorists were then forced to assign preference to the wave over the trajectory, and to give up, in situations like this, any idea of tracking and calculating the trajectory of a particle. This was the basis for the name given by Louis de Broglie to the mechanics that was just coming into being, and would gradually revolutionize 20th century physics, chemistry, and even biology: he called it "wave mechanics," and it would ultimately prove to be the key to understanding and even predicting the properties of atoms and molecules.

The validity of de Broglie's theory was confirmed in 1927 by Clinton J. Davisson and Lester H. Germer. These two physicists demonstrated that, just like a light beam, a beam of electrons could also diffract. This phenomenon thus revealed the existence of the waves which accompany the movement of electrons, just as de Broglie had proposed.

This analogy between a beam of light and a beam of electrons had one consequence of great technological and scientific importance: it led to the creation of the electron microscope, an instrument in which light is replaced by electrons. This new device gave access to magnifications a hundred times greater than those of optical microscopes. The electron microscope revealed extraordinarily tiny details in both inanimate and living matter, and rapidly became an indispensable tool for studying plants, animals, bacteria, and viruses. Today, it can even be used to guess at the outlines of the large molecules found only in living tissue.

NIELS BOHR'S ATOM

The planetary model of the atom had become established thanks to the work of Jean Perrin, Hans Geiger, J. M. Nutall, and Ernest Rutherford. The simplest atom, hydrogen, appeared to consist of a positive nucleus orbited by a single negative electron. It was found that when a little bit of hydrogen at low pressure was sealed in a glass tube, and a high voltage was passed through the gas, the tube would emit a lovely pink light. This was the ancestor of the neon tube so familiar to us today.

There was every reason to believe that emission of photons in this case was the result of excitation of the hydrogen atoms under the action of the electrical discharge. Towards the end of the 19th century, physicists had used a prism to sort out the photons contained in that lovely pink light: they detected red, blue, indigo, and violet photons. It was therefore the superposition of these colors which produced the pink color. One physicist, the Swiss Johann Balmer, also noted that the frequencies f of these four photons were proportional to the very simple expression $1/n^2 - 1/2^2$, where n equaled 3 for red, 4 for blue, 5 for indigo, and 6 for violet.

For a long time, the origin of this curious relationship remained a mystery, but in 1913 Niels Bohr solved the puzzle. He assumed that when an atom emits a photon, what happens is that one of its electrons changes orbit and moves closer to the nucleus. The energy that it loses in the process can be found in the photon that was emitted. With this hypothesis, all that was needed to solve the problem was to look for electron orbits that have energies of the form A/n^2, where A is a constant. And that led Bohr to the idea that in a hydrogen atom, the electrons can revolve around the nucleus only in circular orbits whose radii correspond to energies expressed in the form A/n^2, where n equals 1 for the orbit closest to the nucleus, 2 for the next, and so on. Generalizing this concept with his usual daring, Bohr formulated a new hypothesis which stated that of all the orbits that could be imagined for the electrons of an atom, only a few were possible; and that those few defined the "quantum states" of the atom.

In the case of hydrogen, calculations therefore indicated a radius of .053 nanometers (nm) for the electron orbit closest to the nucleus, in other words the one with the lowest energy. This important result gives some idea of atomic dimensions. In their lowest-energy quantum state (called the "ground" state; all other states are "excited"), ten million of these hydrogen atoms would fit in single file in a space one millimeter wide. Because of Bohr's work, a whole new and vivid language had to be created to describe the structure of atoms. The electron orbit corresponding to $n = 1$ is called the K shell; $n = 2$ is called the L shell, $n = 3$ is the M shell, and so on. It also became necessary to assume that no more than two electrons can be present in the K shell, no more than eight in the L shell, and in general no more than $2 n^2$ electrons in the n shell. This then allowed physicists to construct very precise images of all the quantum states of atoms.

As we shall see, the arrival a few years later of wave mechanics, also called quantum mechanics, would show that these images needed some serious revision.

53

QUANTUM MECHANICS AND THE ATOM

The problem is that in wave mechanics, it is no longer possible to follow in detail, or calculate accurately, the trajectory of a small particle (electron, photon, nucleus, etc.). Quantum mechanics accepts the existence of quantum states which conform to Bohr's concepts, but the way in which the "motion" of electrons in these states is described is very different from the old ways that had to be carried over from classical mechanics.

In classical mechanics, it is possible to calculate the position and velocity of an object at any moment in time; that calculation in turn allows one to determine the object's trajectory with a great deal of accuracy, as is done for a rocket launched into space to explore a distant planet. In wave or quantum mechanics, all that can be done is to calculate the probability of finding a particle at a given moment in a tiny volume surrounding each point in space. The ratio between this probability and the volume is called the "probability density" (or simply the "electron density," in the case of electrons). These probabilities are calculated on the basis of the wave which guides the particles; and quantum mechanics provides, with all the necessary precision, the mathematical formulas needed to make that calculation.

If we now look back at the hydrogen atom, assumed to be in its lowest energy state, we discover that the maximum electron density coincides with the nucleus of the atom. Of all the points in atomic space, it is the nucleus that is most often visited by the hydrogen atom's electron. This observation appears diametrically opposed to Bohr's theory, which depicted the electron orbiting in a circle around the nucleus and therefore never touching it! When two scientific theories contradict each other, the winner is the one that is capable of predicting new phenomena while explaining previous ones: scientists say that they choose the theory with the greatest "heuristic value." In 1935, two Japanese physicists (Hideki Yukawa and S. Sakata) reasoned that if quantum mechanics—the theory constructed by Werner Heisenberg and Erwin Schrödinger in 1927 on the basis of de Broglie's ideas—were the better theory, the nuclei of certain atoms should be able to "swallow" electrons when the electrons brushed by them. This phenomenon was completely impossible according to Bohr's initial theory.

In 1937, two years after this theoretical prediction, Luis Alvarez observed this "electron capture" phenomenon in beryllium. The nuclei of certain beryllium atoms did indeed spontaneously absorb one of their electrons, and turn into lithium. With this process, a gram of the beryllium-7 isotope (^{7}Be) would yield half a gram of lithium in about fifty days, with no outside intervention. This spectacular discovery of a new radioactivity phenomenon demonstrated the superiority of quantum mechanics, and ensured its definitive success. It would also overturn another aspect of the image of atoms that had arisen from Bohr's concepts.

Scientists had acquired the habit of distinguishing among the electrons in an atom, calling the ones in the first orbit "K electrons," those in the second "L electrons," and so on. Since the concept of trajectories or orbits no longer existed in quantum mechanics, this distinction lost all meaning. Quantum mechanics also comprises an important "indistinguishability" principle, which states that all the electrons in a single system are identical, meaning that they cannot be labeled, and that each electron in an atom or molecule plays, on average, the same role in the molecule. In order to retain a geometrical image of an atom, however, I introduced (with the assistance of Simone Odiot), a new concept: that of the "loge."

Consider a helium atom, which has two electrons. According to Bohr's theory, in the ground state the two electrons would orbit in the K shell. In the atom's first excited state, there would be one electron in the K orbit and the second in the L orbit. In quantum mechanics, it is no longer possible to distinguish between a K and L electron; but it is still possible to imagine a sphere of radius R centered at the nucleus of the atom, and to calculate the probability P of finding one electron, and one only, in that sphere. We find that this probability passes through a maximum value of 0.93 (where the value 1 is assigned to an absolutely certain event) when the radius R is one ten-millionth of a millimeter. The corresponding sphere is called the K loge, and the rest of the space is the L loge. We can then state that there is a very high probability of encountering one electron (and only one) in the K loge of a helium atom that is in its first excited state, and consequently a high probability of finding a single electron in its L loge. We are no longer distinguishing between electrons, but between domains of atomic space—loges—in which a constant number of electrons is almost always present.

ABOUT PRECISION

Today, quantum mechanics has become an extremely precise body of mathematical doctrine that can be used to make both qualitative and quantitative interpretations and predictions. Not only has it helped to "mechanize" chemistry, but more generally, a mathematical regime has been imposed upon molecular populations by giving structure to the quantum molecular sciences. These disciplines, since sixty-five years ago in 1927, began to invoke both quantum

57

mechanics and statistical mechanics. To study a given molecular population in the context of this approach, one first uses quantum mechanics to calculate the values characteristic of each molecule constituting the population; then, with statistical mechanics, one predicts the behavior of those molecules as a whole.

Quantum mechanics is essentially based on three principles. Let us consider a set of objects a, b, c, d. What we call the "operator" O for these objects is a mathematical construct which makes each object correspond to another object of the same set. If the operator O makes b correspond to a, c to d, and a to c, we would write: $O \perp a = b$; $O \perp d = c$; $O \perp c = a$. For example, consider the functions $y_1 = x, y_2 = 2x, y_3 = 3x, y_4 = 4x, \ldots$ and an operator O which adds $2x$ to each function. We would then get: $O \perp y_1 = y_1 + 2x = y_3$; $O \perp y_2 = y_2 + 2x = 4x = y_4 \ldots$ Turning our attention now to the equation $O \perp y = n \times y$, we see, for example, that y_1 satisfies the equation, since $O \perp y_1 = x + 2x = 3x = 3y_1$. We say then that y_1 is an "eigenfunction" of the operator O, and that the number 3 is an "eigenvalue" of that operator. Generally speaking, every operator acting on a given set of functions has a series of eigenfunctions and a series of eigenvalues associated with these functions.

This brief mathematics lesson will help us to understand the first principle of quantum mechanics, which can be stated as follows: "Every variable (energy, velocity, position, etc.) that is characteristic of a quantum object (electron, nucleus, atom, etc.) is associated (according to precise rules) with an operator. Every possible measurement of any of these variables corresponds to one of the eigenvalues of the associated operator." Let us assume that we want to know all the possible values for the energy of a molecule of water (H_2O). There are two approaches: either we measure these energies by designing a suitable apparatus; or we calculate the eigenvalues of the operator associated with the energy of that water molecule.

It has been found that in simple cases, the experimental and theoretical approaches lead to exactly the same numbers. However, there are cases in which calculation costs less than experiment, and the obvious choice is to use quantum mechanics. In other instances, experiment is impossible. This situation arises when one wishes to study the properties of a molecule that has been conceived of but not yet synthesized; for example, when one wants to manufacture a medication capable of curing some disease, and an effort is being made to discover a suitable molecule. It is less costly to calculate the properties (using imaginary molecular structures), than to manufacture those molecules and test them to see if they live up to our expectations. Ultimately, of course, the tests will have to be made, but in certain

cases quantum mechanics can suggest which particular molecule should be synthesized.

The second principle is an equation that allows
us to calculate the changes in the wave characterizing
a quantum object when that object is placed under
very specific experimental conditions.

The third principle is used to calculate, from
this wave, the probability of finding one of the eigenvalues of the associated operator when measuring a
characteristic variable of this same quantum object.

QUANTUM IMAGES OF CHEMICAL BONDS

The hydrogen gas used in oxyhydrogen welding
torches consists of very simple molecules resulting
from the union of two hydrogen atoms. This molecule, which contains two hydrogen nuclei and two
electrons, is symbolized as H_2. If one of the two electrons is removed from this molecule, the result is an
even simpler molecule, one that now has only a single electron hovering between the two positive
nuclei. This molecule is therefore no longer electrically neutral, since the single negative electron
cannot compensate for the positive charges of the
two protons. We call this an ionized molecule, in this
case the H_2^+ molecular ion; the "+" sign signifies the
existence of the excess positive charge.

It was perfectly natural to begin the quantum-mechanical study of molecules with this very simple example, and that is what the Danish physicist Carl Burrau did in 1927. His work concentrated especially on the most stable quantum state of the molecule, the lowest-energy or "ground" state. To make the calculations easier, a two-step procedure is used. First the molecular ion is replaced by a model in which the two nuclei are replaced by two fixed positive charges located at an arbitrary distance d. Then the lowest eigenvalue of the operator associated with the energy of this model is calculated in accordance with the first principle of quantum mechanics. This energy turns out to pass through a minimum when the distance d is .106 nm. This distance is called the "equilibrium distance," and the energy at the minimum is equal to 2.7773 electron volts. This molecular model is therefore perfectly stable, and according to the calculations, an energy of 2.7773 electron volts is required in order to dissociate it, in other words remove the positive charges to an infinite distance.

Quantum mechanics tells us that even in this ground state, the nuclei of the real molecule are in perpetual motion, vibrating around the equilibrium distance. Quantum mechanics allows us to calculate the value of this minimum vibration energy, which is called the "zero-point energy." When this energy is taken into account, we find that experimental data relating to the hydrogen molecular ion are in complete quantitative agreement with the results of the quantum calculations. The mean distance between the H_2^+ nuclei, as indicated by experiment, is in fact .106 nm, and the measured dissociation energy turns out to be exactly equal to the value calculated by Burrau.

A new fact of great significance is that even when the molecules are in their lowest-energy state, at a temperature of absolute zero—0 K or –273°C (–460°F)—the nuclei continue to vibrate. It is therefore impossible to halt completely the agitation of the atomic nuclei that constitute ordinary matter.

Burrau's work therefore presents us with a preliminary quantum image of the chemical bond. In its most stable quantum state, the H_2^+ molecule looks like an electron "moving" between two positive nuclei, each of which is vibrating. Since the electron is negative, it attracts the two nuclei towards it; they repel each other, which prevents the molecule from disintegrating. Of course the trajectory of this electron cannot be calculated, but it is possible to calculate from its wave the probability of finding it within any particular small volume of molecular space, and, as in the case of a hydrogen atom, we find that of all the points in that space, the single electron that constitutes the bond visits the molecule's nuclei most often. Chemists sometimes account for this electron by representing it as a dot.

The formula for the hydrogen molecular ion then becomes: $H \cdot H^+$.

In that same year of 1927, three years after the publication of Louis de Broglie's thesis, Walter Heitler and Fritz London used quantum mechanics to study the hydrogen molecule (H_2) itself. This time the problem involved two electrons and two nuclei. It is now time to discuss an important property of particles (electrons, nuclei, photons, etc.): to account for their behavior, scientists have been forced to assume that they rotate, sort of like the Earth rotating about its North–South axis. This property reinforces the analogy between an atom and our planetary system. Associated with every spin is a variable called angular momentum, and the characteristic angular momentum of a particle is usually expressed in terms of a unit equal to $h/2\pi$, where h is Planck's constant.

The characteristic rotational movement of a particle is called "spin." It is customary to measure spin as projected onto a magnetic field, and two very distinct types of particles have been observed as a result: those for which the spin projection value is equal to a whole number multiplied by $h/2\pi$ (which are said to have "integer spin"), and those (like electrons) for which the value is approximately equal to half $h/2\pi$ ("half-integer spin").

Particles with half-integer spin are called "fermions," and they behave in a very specific way. We say that they "obey Fermi-Dirac statistics"; this means that two fermions of the same kind (say, two electrons), which have spin projections of the same sign, cannot occupy the same site. The electrical repulsion that exists between them is therefore joined by a quantum type of repulsion, which is entirely new and much more effective. This quantum repulsion does not necessarily exist for two electrons that have spin projections of opposite signs; in this case only the electrical repulsion applies, and it is much easier to compensate for.

Integer-spin particles, called bosons, obey "Bose-Einstein statistics." No particular quantum repulsion exists between them.

Let us return to Heitler and London's calculations. These physicists used an approach similar to Burrau's, but limited themselves to a more approximate calculation. First of all, they replaced the molecule with a model comprising two electrons and two fixed nuclei separated by a distance d. They found that the lowest-energy quantum state of this model passed through a minimum for a distance d equivalent to .08 nm. The average distance between the nuclei in this state, as indicated by experiment, was .074 nm: the agreement, although not perfect, was still satisfactory.

But now a new and important fact presents itself: in this state, the spin projections of the two electrons have opposite signs. Matters are different in the subsequent quantum state: energy decreases regularly as the distance d increases, and the minimum is therefore obtained when the nuclei are separated by an infinite distance. In this state the molecule is unstable, and the spin projections of the electrons can have the same sign. In the H_2 molecule, therefore, the bond between the two positive nuclei is created by the action of two electrons whose spins have opposite spin projections. These electrons are repelled from each other only by relatively weak electrical forces, thus attracting the two nuclei and ensuring that the molecule is stable. As in the case of the H_2^+ molecule, the nuclei still represent points at which the electron density is highest.

Since this historic first calculation by Heitler and London, many other, more rigorous calculations have been performed. The best of them produce excellent agreement between calculated values for the average distance between the nuclei, and experimentally measured values for those same quantities. Once again considering the electrons in a hydrogen molecule as dots; chemists represent it by the formula H:H.

Let us now look at a slightly more complex molecule, such as the water molecule (H_2O). The oxygen atom contributes eight electrons to this structure, while each hydrogen atom brings one. The H_2O molecule therefore contains 10 electrons orbiting around three nuclei. Chemists have established that only some of these electrons participate directly in the bonds which join the atoms. Following an idea formulated by Gilbert Lewis (1875–1946), they assumed that in this water molecule, two electrons— called valence or bonding electrons—contribute to the stability of each of the two OH bonds. The molecule is then written as H:O:H or H–O–H, where each dash represents a two-electron bond.

This proposition, which is very useful for understanding and even predicting the properties of molecules, does not seem acceptable from a quantum-mechanical standpoint. Since electrons are indistinguishable, it is not "legal" to distinguish between bonding electrons and other ones: each electron must, on average, play the same role in the stability of the molecular structure. But the loge concept can also be applied to molecules.

The simple example of the boron hydride (BH) molecule shows the way. This molecule contains six electrons, five from the boron atom and one from the hydrogen. As soon as the molecule forms, it becomes impossible to distinguish between the electrons from the boron and the electron which belonged to the hydrogen; we are then faced with a problem involving six indistinguishable electrons. But there is nothing to prevent us from abstractly dividing the molecular space into different regions. Quantum mechanics allows us to draw a sphere of radius R around the boron nucleus, and to calculate the probability P of finding in this sphere two electrons, and only two, which have spin projections of opposite signs. This probability P passes through a maximum when R = .04 nm; essentially the same value as in a free boron atom. We then say that this sphere defines the "core loge" of the boron atom in the BH molecule.

If we now define, in the region surrounding this core, a conical segment whose axis is the line joining the boron nucleus to the hydrogen nucleus, we can use the wave associated with the molecule to calculate the probability P of finding two electrons, and only two—which moreover have opposite-sign spin projections—in the conical segment surrounding the BH line. The maximum value for this probability P is obtained when the angle a (at the tip of the cone) is 73°. This position of the cone defines the BH "bonding loge." In the remaining space outside the core loge and the bonding loge, there is a high probability of finding two electrons, and only two. This zone is called the "loan-pair loge."

Thus, using quantum mechanics and the concept of loges, we have divided the molecular space (without distinguishing among electrons) into three loges with two electrons each: the boron core loge, the BH bonding loge, and the loan-pair loge. We can imagine electrons moving throughout the molecular space and visiting each of these loges in turn; the most probable events will correspond to the presence of two electrons in each loge—but not always the same ones. We can also imagine finding, for a brief instant, three or even four electrons in one of the loges, but this is a rather improbable event.

Quantum mechanics is above all a theory which describes phenomena in terms of the calculation of probabilities; it is a probabilistic science. Using the wave associated with a molecule, it is possible to calculate the probability of finding an electron in a small volume surrounding a given point in the molecular space, and thus to prepare a sort of map that indicates, at each point, the value of the electron density in the molecule.

In addition, if we have a crystal made up of identical molecules, we can experimentally determine those same densities by passing a beam of x-rays through the crystal and—using a method that is now commonplace—analyzing the x-ray diffraction phenomena produced by the crystal. In simple cases, when it is possible both to make good quantum-mechanical calculations and to perform a careful diffraction experiment, the calculated results agree with those of the experiment. Here again, one often has the choice between a purely theoretical approach and the experimental route. Both experiment and calculation confirm that the electron density maxima coincide with the positions of the atomic nuclei.

In cooperation with Sylvette Besnaïnou, I introduced a concept that reveals the effect of chemical bonds on the distribution of electron density. The concept is called "differential density"; for any single point, it represents the difference between the density existing at that point in the molecule, and the density which would exist at the same point if the densities of the atoms had simply been superimposed. When this differential density is positive at a certain point, the chemical bond has caused an electron to be attracted to that point. At a point where the density is negative, the bond has led to electron depletion. We then find that in the H_2 molecule, the bond causes electron attraction all along the line segment linking the two nuclei. In the BeF_2 molecule, consisting of one beryllium atom and two fluorine atoms, we find electron attraction on the fluorine side and depletion on the beryllium side.

Pondering this phenomenon in 1947 while working with Wolfgang Pauli's team in Zürich, I arrived at the idea that it must be possible to modify the half-life of certain radioactive nuclei (those that decay by electron capture) by placing them in different molecules where they would not be surrounded by the same electron density.

Emilio Segré, a student of Enrico Fermi's, arrived independently at the same conclusion.

In 1948, along with colleagues at the *Institut du radium*, I compared the mean half-life of radioactive ^{7}Be atoms in beryllium metal with the half-life of identical beryllium atoms in BeF. Since the beryllium nucleus experiences a decrease in electron density in the fluoride, I expected that the half-life of the beryllium isotope in the fluoride would be slightly longer than in the metal. Our experiments confirmed that prediction. For the first time, it became possible to change the rate of radioactive decay of an atomic nucleus. This discovery was confirmed a year later by similar research performed by Segré's team.

This type of experiment, now common, offers a very precise way of measuring the electron density around the nuclei of atoms and molecules.

THE COLOR OF MOLECULES

Molecules absorb light and therefore colors; but they also transmit them. These processes contribute to the beauty of the world in which we live, and also allow us to organize light, for our safety or our pleasure, by using luminescence or phosphorescence. To understand and master these processes, we must once again turn to the quantum molecular sciences.

Let us consider the operator associated with the energy of one isolated molecule in a vacuum, and calculate the eigenvalues of that operator. The first principle of quantum mechanics teaches us that each number that we obtain corresponds to one possible energy for the molecule. The wave associated with each energy level that we then find corresponds to a quantum state of the molecule, in other words a complex motion of the electrons and nuclei, described in terms of probability.

By way of first approximation, we can dissect this motion into four elements. First, the motion of the molecule's center of gravity, whose energy depends on velocity: this is called the molecule's kinetic energy. Second, the motion of the electrons with respect to the nuclei, corresponding to a certain electronic energy. Third, the vibratory motion of the nuclei (vibrational energy). Fourth and last, the rotating motion of the nuclei (rotational energy). We can already guess that the molecule must have a very large number of possible energy levels.

If we now immerse our molecule in a beam of light (or, more generally, a bath of photons), it will absorb a photon if the photon's energy is identical to the difference between the energy of the level at which the molecule is now, and the energy of another level. If the photon's energy is low (far infrared), it will only be able to change the rotational motion of the nuclei; we say then that the photon belongs to the molecule's "pure rotational" spectrum. If the photon has a little more energy (near infrared), it may be able to change both the rotation and the vibration of the nuclei. The photon then belongs to the vibrational–rotational spectrum. Lastly, in the case of a photon in the visible-light or ultraviolet region, it will be capable of modifying electron motion. We then say that this photon is part of the molecule's electronic spectrum.

Conversely, a molecule can drop from a given energy level to a lower level, in which case it emits a photon. These phenomena, in which molecules absorb and emit light, are responsible for the light-related effects produced by the objects that surround us. The radiation emitted by the Sun is a complex mixture of very diverse photons whose frequencies extend essentially from the infrared to the ultraviolet, containing all the colors of the rainbow.

If we dissolve a little iodine in ordinary alcohol, we obtain a solution similar to the tincture of iodine used in medicine. Now let us pour a little of this solution into a glass container shaped like a square-based prism, and shine a beam of sunlight onto one of the faces of the container. Looking in the direction opposite to this incident light, we see that the light which emerges from the glass—the light transmitted by the solution—is brown. Since we know that brown is not a pure (or spectral) color, the transmitted light must therefore consist of a mixture of photons of different colors, but one that is different from the mixture constituting sunlight, which looks white. We can surmise that the iodine molecules have absorbed photons corresponding to certain energy level differences, while allowing others (the components of the brown light) to pass.

The molecules absorb photons which, as a group, correspond to a certain color (the "absorbed color"), and pass on to our eyes other photons which represent the "transmitted color." The table below lists certain transmitted colors corresponding to the colors that are absorbed when a molecule is illuminated with white light.

ABSORBED COLOR	TRANSMITTED COLOR
Violet	Green-yellow
Blue	Yellow
Green-blue	Red
Green	Purple
Orange	Blue-green
Red	Green-blue

We can now see why the color of the light transmitted by the molecules depends on the color of the light with which they are illuminated. Consider a molecule that absorbs photons of a certain shade of red. If a material consisting of those particular molecules is illuminated with a red light containing only those photons, it will absorb all the photons and allow nothing to pass. The material will look black. If we now illuminate it with a yellow light that is not absorbed by the molecules, it will look yellow.

But what happens to the energy of the photons absorbed by the molecules? Remember that a specimen of gaseous, liquid, or solid matter consists of a huge number of molecules: a molecular population usually comprises billions of billions of individuals. At room temperature, these molecules are often at their lowest electronic energy level, but at various vibrational–rotational levels.

We also know that in a gas, for example, these molecules, traveling at high speeds, are constantly colliding with each other—what we call "thermal agitation." If the photon absorbed by a molecule of this population belongs to the infrared region, its energy will change (or more precisely, will amplify) only the vibrational–rotational motion of the molecule. As it experiences collisions, the molecule will then share this excess energy with its neighbors; and the thermal agitation—and therefore the temperature—of the gas will then have increased very slightly.

Similar phenomena occur in liquids and solids. When the photon is in the visible or ultraviolet range, it changes the electron motion in the molecule that absorbs it. Here again, in the course of the billions of collisions that subsequently occur, the photon's energy can disperse by increasing the thermal agitation; that is, it can dissipate by creating heat. But sometimes another phenomenon occurs: the molecule can fall back, from the excited electronic level to which it was raised by absorbing the photon, to a lower electronic level, and in the process emit another photon. It has thus captured a photon of a certain energy or color, and emits another photon which often has a slightly lower energy and is therefore a different color. If emission occurs almost immediately after absorption, we witness the familiar phenomenon of fluorescence. If emission is delayed for a certain period, we speak of phosphorescence.

Let us take the example of how uranium salts behave under ultraviolet light: they emit a pretty yellow-green light which disappears as soon as the ultraviolet light is turned off. This is then a fluorescence phenomenon. Since ultraviolet light is invisible to our eyes, if we perform the experiment in a dark room, we see the uranium salts shining in the darkness. This is the principle which underlies the spectacular "black light" effects sometimes created in theaters and nightclubs.

Since quantum mechanics in principle allows us to calculate the energy levels of a molecule, it can provide, even for an imaginary molecule, some very valuable information that can be used to predict its color, its fluorescent properties, and even its phosphorescent effects. Quantum mechanics can thus guide the chemist who is trying to prepare new dye materials.

A QUANTUM THEORY OF CHEMICAL REACTIONS

Earlier we described the combustion of sulfur: sulfur atoms combine with oxygen in the air, producing molecules of sulfur dioxide (SO_2). These combustion reactions, of which there are many, release heat, and the resulting hot gases constitute a flame. Oxyhydrogen welding torches burn a mixture of hydrogen and oxygen, yielding water vapor. This reaction is represented by the chemical equation $2H_2 + O_2 \rightarrow 2H_2O$, which can be read as follows: two hydrogen molecules react with one oxygen molecule to form two water molecules. When ordinary cooking gas burns in air, oxygen combines with the hydrocarbon molecules (which consist of hydrogen and carbon atoms) in the gas, forming mostly water and some carbon dioxide, whose molecules contain one carbon atom and two oxygen atoms.

The little explosions that make a car engine run also involve this kind of reaction. The name "oxidation" is also applied to all these reactions produced by oxygen; and by extension we can say (as we already have) that our cells use the respiratory chain to "burn" the materials derived from what we eat. The protein synthesis that is constantly taking place in our cells is another example of a chemical reaction that occurs in living tissue. These are sometimes called "biochemical" reactions.

There is no essential difference between the chemical reactions that occur in inanimate matter, and biochemical reactions. The latter, however, often utilize much larger molecules, and their mechanisms are more complex. Certain reactions require the presence of radiation, and these are called "radiochemical" reactions. If the radiation consists of photons, the corresponding reactions are called "photochemical" reactions.

Let us imagine that in a darkened room, we experiment by enclosing a mixture of chlorine and hydrogen in a transparent plastic bag. (Chlorine is a greenish-yellow gas with a suffocating odor, consisting of diatomic Cl_2 molecules, where Cl is the symbol representing one chlorine atom.) As long as the room remains dark, nothing happens. But if a sunbeam touches the plastic bag, it will explode: heat will be released as the Cl_2 molecules combine with the H_2 molecules to produce molecules of hydrochloric acid (HCl). The chemical equation representing this reaction can therefore be written: $Cl_2 + H_2 \rightarrow 2\ HCl$.

But this equation does not claim to account for the mechanism by which this reaction occurs. Detailed studies have shown that it is, in fact, a chain reaction. First a chlorine molecule absorbs a photon, which splits the molecule into two chlorine atoms: Cl_2 + photon $\rightarrow$ Cl + Cl. Each resulting atom of free chlorine can combine with one hydrogen molecule to form one molecule of hydrochloric acid and one hydrogen atom: $Cl + H_2 \rightarrow HCl + H$. Each hydrogen atom produced by this reaction can in turn combine with one chlorine molecule to form one molecule of hydrochloric acid and one chlorine atom: $H + Cl_2 \rightarrow HCl + Cl$, and the regenerated chlorine can continue the chain almost indefinitely. In fact, one single photon is sufficient to produce 100,000 molecules of hydrochloric acid, which is why the reaction is explosive. The chain can only be broken by secondary events (impurities, collisions with the container walls, etc.).

The synthesis of chlorophyll by plants and the reactions produced in the retina of our eye by light show very clearly the importance of photochemical reactions in living matter. These reactions allow us to see the world around us; and plants, which are eaten by animals, are themselves constituted by photochemical reactions.

The example of the reaction between chlorine and hydrogen leads me to emphasize once again the fundamental difference between the equation $Cl_2 + H_2 \rightarrow 2HCl$, which simply records what is called an "analytical balance" (the result of analyzing the reaction product), and the following equations:

$$Cl_2 + photon \rightarrow Cl + Cl$$
$$Cl + H_2 \rightarrow HCl + H$$
$$H + Cl_2 \rightarrow HCl + Cl$$
$$Cl + H_2 \rightarrow HCl + H$$

and the like, which represent the actual sequence of elementary processes that constitute the reaction, and illustrate its subtle mechanism.

Quantum chemistry primarily involves studying in depth the mechanisms of chemical reactions with the simultaneous use of quantum mechanics and statistical mechanics. An investigation of this kind is generally possible only when a previous experimental study has already provided a certain amount of information about the probable mechanism of the reaction.

When two molecules collide, and their velocity is great enough, a reaction occurs between them. The most elegant way of bringing them into contact is to use the intersecting molecular beam method. Doing so requires a large chamber in which a very good vacuum can be created, so as to suppress undesirable reactions. Equipment capable of producing beams of molecules traveling at known speeds is then installed in the chamber.

Let us suppose, for example, that we wish to undertake a detailed study of the interaction between bromine and the metal potassium. One of the molecular beam generators will contain a little bit of liquid bromine, whose diatomic Br_2 molecules (Br being the symbol for the bromine atom) will emerge through a small orifice and form a very thin beam. The second generator will contain a little heated potassium metal, which will produce a beam of K atoms (K being the symbol for potassium, derived from the Latin *kalium*). We then set up the two generators so that the two beams meet at the center of the chamber, forming a preselected angle *a* between them. By moving one of the generators with respect to the other, we can change the angle *a*; and by heating the bromine and potassium to various temperatures, we can precisely adjust the velocities of the molecules in the beams. The reaction which occurs when the two beams meet is very simple; it can be written: $K + Br_2 \rightarrow KBr + Br$; in other words, one potassium atom colliding with one bromine molecule yields one molecule of potassium bromide (KBr), and one bromine atom.

The chamber contains other instruments called "chemical detectors." In the present case, one of these detectors consists of a strip of tungsten capable of stripping an electron away from any potassium

atom or KBr molecule that touches it. When connected to an electron counter, this detector accurately
determines the total number of K atoms and KBr
molecules that strike it during a certain time interval.

The second detector consists of a strip of a tung-
sten/platinum alloy containing 8% tungsten. This
alloy can still capture an electron from potassium,
but is insensitive to KBr molecules. The detector
therefore counts potassium atoms. By subtracting its
count from the number provided by the first detector,
we find out how many KBr molecules were formed.

The high-performance apparatus that we have
just described can thus accurately measure the number of Br atoms formed during the collisions that
occur between Br_2 molecules and K atoms that meet
at a specific angle and at known speeds. Conditions
turn out to be excellent for obtaining plentiful and
accurate experimental information. We therefore find
that if the collisions occur between potassium at
413°C (775°F) and bromine at 41°C (106°F), the
angle most favorable to the formation of potassium
bromide molecules is 30 degrees.

Using quantum chemistry, we can calculate the
probability that two molecules will react when they
meet at a certain angle and at certain speeds. We first
use the second principle of quantum mechanics,
namely the equation which allows us to calculate the
change over time in the wave associated with a sys-

tem of particles. In the case of our potassium/ bromine reaction (K + Br$_2$ $\rightarrow$ KBr + Br), we first look at the wave associated with the system consisting of one potassium atom and one bromine molecule, with the two initially located very far away from each other; then, using the wave equation, we examine how the wave changes as the atom approaches the molecule, and finally as they collide. By observing this wave, we can then calculate the probability that after the collision, the K + Br$_2$ system will become the KBr + Br system. We have thus determined the probability that a reaction will take place, and in principle we can then calculate how that probability changes as a function of the angle between the particle trajectories at the moment of impact. But although these calculations are theoretically easy to describe, they are extremely laborious to actually perform, even with powerful computers.

So far, highly accurate calculations have been possible only for very simple reactions. One such reaction is D + H$_2$ $\rightarrow$ DH + H, where the symbol D represents one atom of deuterium, or heavy hydrogen. Let us digress briefly and remember that a given element, hydrogen for example, can exist in a variety of atomic forms. This fact has to do with the structure of atomic nuclei, which are made up of two types of particles—protons and neutrons—with very similar masses.

To simplify matters, we should remember also that protons have a positive charge which is equal in magnitude to that of the electron, but of opposite sign. Neutrons, as their name suggests, are electrically neutral; they have no electric charge. Most hydrogen atoms have a very simple nucleus consisting of a single proton and a single electron, which orbits around the proton and makes the atom neutral (zero total charge). This is called "light hydrogen," or simply H. But there are certain much rarer hydrogen atoms in which the nucleus contains a proton *and* a neutron, which are orbited by a single electron. These hydrogen atoms therefore weigh approximately twice as much as ordinary H, and hence are called "heavy" hydrogen or deuterium. We can also explain the relationship by saying that the H and D atoms are two different isotopes of a single element (hydrogen).

Back to our reaction: $D + H_2 \rightarrow DH + H$, which is highly suitable for both intersecting molecular beam experiments and quantum mechanical calculations using the collision theory that we have just discussed in outline. Calculations performed in 1965 in the United States by M. Karplus, R. N. Porter, and R. D. Sharma proved remarkably consistent with experimental results: they provided information about events that occur when the deuterium atom collides with the hydrogen molecule. We know that the nuclei of this molecule possess vibrational/rotational motion. At the moment of collision, and for an unimaginably short period of time (ten *trillionths* of a second), a triatomic molecule forms; this DHH molecule is sometimes called an intermediate complex.

Two phenomena can then occur: either the complex breaks up and re-forms into its initial partners (DHH → D + H_2), or the deuterium and one of the light hydrogen atoms form a DH molecule whose nuclei begin vibrating and rotating, while the other light hydrogen atom goes off in some other direction. The result of the collision can then be written: DHH → DH + H; a reaction has taken place.

Quantum mechanics allows us to calculate the probability that one or the other of these events will occur, given certain experimental conditions. The calculations reveal that the reaction rate is an exponential function of the inverse of the absolute temperature at which the reaction takes place. In other words, the rate doubles every time the inverse of the temperature decreases by a certain amount. To put it even more simply, the rate rises very rapidly as the temperature increases, as predicted by an equation governing a variety of chemical reactions, discovered many years ago by the Swedish chemist Svante Arrhenius (1859–1927).

In principle, then, quantum chemistry provides a way to interpret chemical reactions, and even to predict reactions not yet observed experimentally—including reactions between imaginary molecules that have not yet been synthesized. Quantum chemistry can therefore, again in principle, help guide the chemist who is looking for new ways to synthesize known molecules or new ones. But because the collision theory entails highly complex calculations, and because the intersecting molecular beam method is not an economically feasible synthesis technique, some changes have had to be made in quantum chemistry, which has gained in usefulness what it has inevitably lost in theoretical rigorousness.

A QUANTUM THEORY OF INTERMOLECULAR FORCES

The chemical reactions performed in industry involve gases, liquids, solids, and sometimes cells. In every case, they act on complex molecular populations. Take the case of a gas at room temperature, in the absence of radiation. Almost all the molecules will be at their lowest-energy electronic state, but some will have various levels of vibrational/rotational energy. In technical language, we say that these levels have different populations. We also know that each of these molecules, traveling at enormous speeds (several hundred meters per second), collides with its neighbors billions of times a second. Just as in the intersecting molecular beam chamber, these collisions can produce reactions. As soon as the temperature of the gas is increased, this thermal agitation also increases, and the population corresponding to the higher vibrational/rotational levels becomes more numerous. Reaction rates therefore rise. In addition, forces are constantly coming into being among the molecules of this gas. Consider two molecules separated by a great distance, that is, a distance greater than their own dimensions. The forces that are exerted between them are called long-range forces; they are always attractive. In other words, when two molecules gaze upon each other from afar, they are always attracted to one another. This attraction is attributable to three types of forces: electrostatic forces, which result from the attraction exerted by the electrons in one molecule on the nuclei of the other; polarization forces, caused by the fact that one molecule changes the electron density distribution in the other; and dispersion forces, arising from interactions between the motions of the electrons in the two molecules.

When the molecules come very close together, the attraction usually turns into repulsion: when two molecules get a close look at each other, they generally push one another away. We then talk about short-range forces. This repulsion essentially results from the fact that two electrons with the same spin projection cannot coexist in a small volume of space. These repulsive forces can, however, be partly compensated for if the molecules in question have specific structural elements. For example, if one of the molecules has a hydrogen atom whose positive proton can attract the electrons in a loan-pair loge belonging to another molecule, the two molecules can link up. The result is what we call a "hydrogen bond."

These hydrogen bonds exist between water molecules, especially in liquid or solid water (ice), since the same types of forces are at work in these two states of matter. In both these states, however, the molecules are very close to one another; one might even say that they are touching, and so the short-range forces become particularly significant. That is why liquids and solids are relatively incompressible.

The innumerable collisions that take place among the molecules of a gas, a liquid, or a solid can, as we have seen, make it possible to perform chemical reactions. This is because the speed at which the molecules collide (their "kinetic energy") can overcome the repulsive forces that always arise in such situations. In technical language, we say that the kinetic energy is capable of overcoming or exceeding the energy of the "potential barrier" that stands in the way of the reaction. It should therefore be obvious that the lower this potential barrier, the more readily the reaction will occur. The higher the barrier, the more the molecules will need to be heated before they react.

THE REACTIVITY OF CONJUGATED MOLECULES

Reaction rate calculations using the quantum theory of collisions are long and complex, and consequently very costly. Quantum chemistry, however, makes it fairly easy to evaluate potential barriers for certain reactions and certain families of molecules. The case of conjugated molecules is worth closer consideration. When a molecule contains only two atoms, the bonding loge necessarily extends between the core loges of those two atoms. But when a molecule is the result of a combination of a larger number of atoms, two types of bonding loges can be identified within the molecule: those which extend only between two adjacent atomic cores (which are called "localized bonding loges"), and those which touch more than two atomic cores (called "delocalized bonding loges"). The molecules referred to by chemists as "conjugated molecules" all contain delocalized bonding loges. This is a very important family of molecules, since it includes dyes, antibiotics, and substances capable of inducing cancer.

88

First let us consider the conjugated hydrocarbons (which consist solely of carbon and hydrogen atoms), specifically the condensed aromatic hydrocarbons. The leader of this tribe is benzene, a familiar solvent. The benzene molecule has an elegant edifice, containing just one hexagonal cycle made up of six carbon atoms, each linked to a hydrogen atom. A two-electron bonding loge is located between each pair of adjacent carbon cores. The conjugated nature of this molecule results from the additional presence of a six-electron bonding loge that extends around the set of six carbon core loges. Molecules belonging to this series include naphthalene (otherwise known as moth flakes), which contains two hexagons whose vertices consist of carbon cores. These two hexagons are linked together by a common side. Anthracene has three linked carbon hexagons, and pyrene has four, arranged in the shape of a cross. Note also that all these molecules are flat (by which we mean that all the carbon and hydrogen nuclei vibrate around the same plane).

Quantum chemistry offers a way of predicting the chemical reactivity of these molecules. We will look at the example of nitration. The reagent usually used to produce nitrated derivatives of these conjugated molecules is a mixture of sulfuric acid and nitric acid. Treating benzene with this mixture readily yields mononitrobenzene (mirbane oil), a liquid with a pleasant odor of bitter almonds.

Two teams of physical chemists, one led by Sir Christopher Ingold in England, the other by Chédin in France, have shown that the active agent in the acid mixture is a positive molecular ion symbolized $NO_2{}^+$, which therefore contains one nitrogen atom (N), and two oxygen atoms. This molecule is called "nitronium." When pyrene is treated with the sulfuric/nitric mixture, how do we know to which carbon atom the nitronium ion will preferentially attach itself? A chemist could never find out simply by looking at the formula, since all the carbon cores linked to a hydrogen atom look the same, and one would expect the nitronium ion to attach to any of them with the same ease. In this instance, quantum chemistry provides a precise answer by simply calculating the energy of the potential barrier involved in this reaction.

Let us consider the case of pyrene, the cross-shaped conjugated molecule. Calculation shows that the carbon cores corresponding to the lowest potential barrier are adjacent to the ones forming the two ends of the long arms of the cross. Theory therefore predicts that these cores should be the most reactive ones—and that is indeed where the reaction occurs.

The predictive ability of the theory seems so good that when experimental results do not agree with theoretical predictions, one is tempted to doubt the accuracy of the experiments. For example, M. J. S. Dewar and E. W. Warford calculated the potential barriers involved in the nitration of phenanthrene. They found that the results of their calculations were not consistent with experimental measurements performed earlier by Schmidt and Heinle. Dewar and Warford then repeated those experiments using a more refined method than their predecessors, and the new results provided perfect confirmation of the theoretical calculations.

X-ray diffraction experiments have taught us that the average distances separating the adjacent carbon nuclei in a conjugated molecule are not identical. Quantum mechanical calculations allow us to understand the reason for those differences. They are essentially due to the existence of the delocalized bond, which makes no contribution to equalizing the distances between the various pairs of adjacent carbon nuclei. In chemical language, we say that it introduces different "bond energies," and this energy can be expressed as a number called the bonding coefficient. This coefficient increases as the distance between the two nuclei decreases: the bond energy helps bring the nuclei closer.

The concept of the bonding coefficient was introduced by the English chemist Charles Coulson in 1939. In 1945, working with Alberte Pullman, I introduced another useful coefficient: the free valence coefficient. This is a quantum indicator of what chemists call the "residual affinity" of an atom that is part of a molecule. To calculate the free valence coefficient characterizing a carbon core in a conjugated molecule, the sum of the bonding coefficients associated with the adjacent carbon core pairs sharing that atom is subtracted from a fixed but arbitrarily selected number. In 1950, along with my colleagues Sandorfy, Vroelant, Yvan, and Chalvet, I demonstrated that as this coefficient increases, the corresponding potential barrier becomes lower, and the atom therefore becomes more reactive.

This is a fundamental correlation, which establishes a link between the reactivity of the atoms in a molecule and the distribution of the electron density in that same molecule (in other words, its structure).

We still need to look briefly at the quantum theory of photochemical reactions. We know that in atoms, the "motion" of the electrons changes when their energy changes, that is, when they move from one quantum level to another. The same is true for molecules, when one compares two different electron levels.

Starting in 1946, again in collaboration with Alberte Pullman, I studied the butadiene molecule, a conjugated hydrocarbon containing four carbon atoms and six hydrogen atoms. Inside this molecule in its ground state, the carbon nuclei form a trapezoid with three short sides and one long side (which therefore corresponds to a low bonding coefficient). If the molecule absorbs an ultraviolet photon which raises it to its first excited state, we find that this bond index increases sharply.

For some time, this theoretical prediction had no application. But seventeen years later, in 1963, B. Srinivasan showed that if a solution of butadiene in ether is exposed to ultraviolet radiation, that bond does indeed become stronger, and cyclobutene is produced. Since then, quantum chemistry has yielded predictions or interpretations of a number of photochemical reactions.

Here is one more important example: using fairly simple calculations based on quantum mechanics, Joseph Dannenberg predicted the possibility of photochemically converting quinodimethane into propellane, a molecule whose shape is reminiscent of a ship's propeller. He then confirmed the validity of this theoretical prediction by experiment. This was a significant result, since propellane until then had been difficult to prepare and expensive to synthesize, while quinodimethane is an inexpensive molecule. The synthesis method revealed by quantum chemistry therefore lowered the cost of synthesizing this particular molecule.

III

THE QUANTUM

MOLECULAR

SCIENCES

IN BIOLOGY

AND MEDICINE

THE STRUCTURE OF PROTEINS AND NUCLEIC ACIDS

The general public is often unaware of the relationship between quanta and the development of the life
sciences. This is regrettable, since biology offers an
immense field of application of the utmost interest
for the quantum molecular sciences. We know that
quantum mechanics offers the possibility of deter-
mining, in principle and by pure calculation, the
distances that separate the average positions of the
nuclei in the constituent atoms of a molecule when
that molecule is in one of its quantum states. The
method used in the case of the H_2^+ molecular ion can
be applied generally. The molecule is therefore

replaced by a "fixed nuclei" model, and the arbitrary distances separating those nuclei are varied to produce the minimum energy corresponding to the level in question.

In an interesting paper dealing with molecular modeling, Bernard Pullman recently summarized the role played by these models in the discovery of the structure of proteins and nucleic acids. In the 1950s, two teams were conducting research on the structure of proteins. One was led by the famous quantum chemist Linus Pauling in California; the other, by three no less famous physicists at Cambridge: William Bragg, Max Perutz, and John Kendrew. It was known that proteins produced by the linkage of amino acids were constructed on the basis of a system formed by the three atoms N, C, and O, which was called the "peptide bond." Pauling also knew that these atoms were joined by a delocalized bond, which he believed were located in a single plane; the British were unaware of this. That is why Pauling's team discovered the helical structure of proteins and received the Nobel Prize.

A little later, following Pauling's example, James Watson and Francis Crick would demonstrate the helical structure of nucleic acid and be awarded another Nobel Prize.

MEMBRANE RECEPTORS AND BRAIN FUNCTION

The starting gun for modern research on the detailed structure of our nervous system was fired by Santiago Ramón y Cajal at the beginning of this century. This Spanish scientist had shown that the essential cells—called neurons—making up our brain, cerebellum, and nerves have very specific forms. They consist of a star-shaped body which contains the nucleus; extending from this star-shaped body are a long tube called the axon, and sinuous, branching filaments called dendrites. Certain kinds of electrical currents (the nerve impulses emitted by the star-shaped body) circulate in the axon and the dendrites, and jump from neuron to neuron across the tiny spaces that separate them (called synapses).

These currents accompany our thoughts, and allow us to see, feel, hear, and give orders to our muscles to act. It is obviously very important to analyze in some depth all these phenomena, which lie at the very heart of our existence. The quantum molecular sciences, which provide precise information about the nature of molecules and the kinds of interactions that can take place between them, constitute one of the essential tools for research in this fascinating field.

To understand the nerve impulse, we begin with salt. Common salt is sodium chloride; its formula is NaCl, where Na stands for an atom of sodium (*natrium* in Latin). When common salt dissolves in water, it produces negative Cl⁻ ions and positive Na⁺ ions, meaning that an electron leaves the sodium and attaches itself to the chlorine. The human body contains common salt (which is present in our food) and lots of water, and is therefore rich in Cl⁻ and Na⁺ ions. These ions permeate our neurons and the spaces that separate them, in which we also find a strange protein called ATPase, which is capable of passing through the membrane that surrounds each neuron. This protein looks for Na⁺ sodium ions inside the neurons, and transports them to the outside; in other words, it acts as a "pump" for the Na⁺ ions. As a result, the interior of the neurons is negative, while the exterior is positive. Each neuron thus is like a tiny electric battery, with a negative terminal on the inside and a positive terminal on the outside.

At rest, the neuron does not allow the passage of the sodium ions that have accumulated on the outside. But if it receives an electrical pulse exceeding a certain value, or threshold, it allows a certain quantity of the ions to enter. The result is then a small electrical current, formed by these positive ions passing through the membrane, which propagates step by step along the entire axon: this current constitutes the nerve impulse.

In humans, this impulse propagates at a speed of about 100 meters per second, which is why our reflexes are so fast. In jellyfish, impulses travel at less than 10 *centimeters* per second (one of many reasons why a jellyfish would never pass a driving test!).

What happens when the impulse arrives at the end of a neuron and stops at the cleft (synapse) that separates it from the next neuron? For the moment, we will discuss only chemical synapses. In the vicinity of these synapses, at the end of the axon, are little reservoirs called vesicles, which contain specific molecules known as neurotransmitters. When the nerve impulse arrives at the end of the axon, some of these vesicles open, releasing millions of neurotransmitter molecules into the synapse.

On every cell membrane there exist molecules, usually proteins, that are capable of recognizing and attaching to certain other molecules. These special membrane molecules are called receptors. Neuron membranes happen to contain specific receptors for neurotransmitters. As they leave the end of an axon and cross the synapse, the neurotransmitter molecules therefore come into contact with the receptors of adjacent neurons, thus transmitting the nerve impulse. To gain some idea of the complexity of events inside our brain, consider that one cubic millimeter of nerve tissue contains approximately 600 million synapses. Each neuron is connected to huge numbers of adjacent neurons by huge numbers of synapses. It continually receives a flow of nerve impulses, and "decides," on the basis of those impulses, which impulses of its own to transmit in turn.

A detailed study of the interaction between a membrane receptor and a neurotransmitter molecule thus represents a pivotal element in accurately understanding how the human brain functions. It is difficult, however, to isolate these receptors and determine their structure experimentally. The methods of the quantum molecular sciences, on the other hand, allow us to determine accurately the electronic structure of the neurotransmitter molecules, which are fairly simple. Moreover, having defined the exact nature of these structures, we can make certain inferences about the structure of the receptors. Seeing that one part of the transmitter molecule has an excess of electron density, for example, we might conclude that the receptor must have a deficit of electron density, so that electrostatic forces will create an attraction between the receptor and the neurotransmitter molecule. Some have remarked ironically that this method is as risky as trying to guess someone's hair color by looking at his or her dog! Nevertheless, some encouraging results have already been obtained with this approach.

In any event, what we have just said already gives us a better understanding of the mechanism underlying our sense organs. We know, for example, that the human eye operates sort of like a camera: its lens acts just like a camera lens, forming (on the retina) tiny images of objects that we look at. What still needs to be explained is how photons, striking the retina and forming the little image, can trigger nerve impulses that pass through the optic nerve to our brain and create the impression of light.

Here is where the quantum molecular sciences can make an indispensable contribution. Our retina contains two types of cells which can convert photons into electrical impulses: cones and rods. The cones, which are responsible for sensations of color, are divided into three categories depending on the frequency range to which they are sensitive. The particular color that is perceived results from a comparison of the signals created by illumination of these three types of cones. Assume, for example, that the first category of cones turns out to be sensitive primarily to red, the second category to yellow, and the third to blue; if only the first category of cones reacts, we will always see red. If the first two categories of cones are stimulated, we will see a shade of orange that will get closer to yellow as the stimulated cones include more and more of the second category. But the cones are not very sensitive: they are stimulated only by a large number of photons, which means a fairly intense image.

The rods, which are more sensitive, do not distinguish colors. This explains why all colors look alike in dim light: it is, in fact, true that "at night, all cats are grey." Cones and rods operate on the same principle. The rods, for example, are cylindrical cells packed together on end over the entire retina, like drinking straws in a box. Approximately every other photon that strikes them triggers a signal that reaches the brain.

Two stages have been identified in this remarkable phenomenon. The first, "transduction," involves absorption of the photon and production of an initial cell signal. In the second phase, this information is processed by the brain. Transduction occurs in the upper part of the rods, inside which is a membrane that contains very specific molecules, including molecules of "retinal" linked to molecules of a protein called rhodopsin. The retinal molecule, like almost every molecule, is capable of absorbing a photon of visible light, which raises the molecule from its ground state to an excited electronic energy level.

Quantum mechanics tells us (and experiment confirms) that this changes the motion of the electron as well as the motion of the molecule's nuclei, and that consequently the average distances separating the nuclei become different, so that finally the shape of the molecule itself undergoes a transformation. In the case of retinal, that change has an effect on the rhodopsin molecule, which thereupon gives the order to manufacture a neurotransmitter-like molecule called guanine nucleotide (namely, 3'- or 5'-guanylic acid). This neurotransmitter molecule then attaches to a receptor on the rod's external membrane. It opens the way for the surrounding sodium ions, which rush into the interior of the rod and create an electric current. This current propagates to the bottom of the rod where the synapses are located, just like a nerve impulse traveling along an axon.

Such is the explanation for the marvelous phenomenon of transduction which allows us to see the world around us; now we can understand how a single photon can trigger the departure of a nerve impulse from a rod or a cone. These nerve impulses that travel from the retina along the optic nerves enter the brain in a specialized region called the area striata or striate cortex. The information collected there is interpreted in adjacent regions, which is how we develop our perceptions of light. The way in which our brains use similar pathways to issue orders to our muscles can also be explained, and we have even managed to describe the complex molecular and electrical phenomena that take place in our brains when they experience the emotions of love, affection, ecstasy, and disappointment.

THE ACTION OF ANTIBIOTICS

The advent of antibiotics, following the discovery of penicillin by the English bacteriologist Alexander

Fleming, has literally revolutionized the treatment of infectious diseases. We know that antibiotics halt the proliferation of bacteria by preventing the synthesis of molecules essential to these parasites' survival. An increasing number of antibiotics has now been synthesized to aid in the fight against the extremely wide range of diseases caused by bacteria. Once again, the quantum molecular sciences are making an important contribution to the investigation of the mechanisms involved.

One nice example is the research performed on tetracyclines in 1971 by the Spanish scientist F. Peradejordi, and A. Martin and A. Cammerata in the United States. These molecules contain carbon atoms linked together to form four hexagonal cycles (*tetra* being the Greek word for "four"). A portion of the molecule is conjugated and flat, and contains a delocalized bond loge. This group of four cycles can be supplemented with a variety of atoms, bonded to specific carbon atoms, to produce a whole series of compounds. The chemist's task is to guess which added atoms will produce the most active antibiotic for a particular type of bacteria.

Experimental studies have helped identify the mechanism of action of the tetracyclines, and have revealed that these antibiotics interfere with synthesis of the proteins that the bacterium needs in order to live and reproduce. We know that the message triggering synthesis of a protein is carried by a gene, which consists of a segment of the deoxyribonucleic acid (DNA) contained in the cell's nucleus. This gene is first copied in the form of another molecule having the structure of a ribonucleic acid, which we abbreviate RNA. This process is referred to as transcription. Then each resulting molecule of RNA attaches itself to a little organelle, present in each cell and called the ribosome, which contains about a hundred large molecules. These ribosomes are essential components of the cell's machinery. The "messenger RNA" attaches itself to a ribosome, and moves past it little by little.

When a group of three RNA bases called a "codon" appears at a particular point on the ribosome, another molecule called "transfer RNA," or tRNA, positions itself on the ribosome opposite the codon. This tRNA molecule is bound to the amino acid corresponding to the codon. As the messenger RNA molecule passes by, the ribosome gathers in the various amino acids corresponding to the various codons, and therefore to the message contained in the RNA molecule. Other molecules weld these amino acids together and create the protein encoded in the messenger RNA, and ultimately in the gene of which it is a copy. This complex phase of protein synthesis is called translation.

But it has been found that tetracyclines attach to the ribosome at exactly the same site as the tRNA molecules. If a sufficiently large dose of tetracycline is sent into a bacterium, the molecules of this substance will attach to the bacterium's ribosomes and, if they attach securely, they will prevent synthesis of the proteins that the bacterium needs. In this respect, tetracycline deserves its designation as an "antibiotic," in other words a molecule which suppresses life. And yet, the antibiotic must not interfere with life in our own cells. We are in luck, however: our ribosomes are very different from bacterial ribosomes. Tetracyclines do not attach to ours, and as a result are not toxic at antibiotic doses. All that remains is to choose the best tetracyclines, and here is where the quantum molecular sciences again can help.

In order for a tetracycline to be highly active, it must first of all be capable of penetrating easily into the bacterium; secondly, it must attach securely to the bacterial ribosome. At present, only experiment can tell us about the first characteristic, but quantum mechanics can provide us with data about the second. The problem, after all, involves the interaction between two molecules: tetracycline on the one hand, and the ribosome receptor on the other. The structure of the antibiotic is known, and analysis of the wave associated with the simple formula of the tetracycline molecule (even if it does not yet exist) allows us to calculate its reactivity coefficients (along with bonding coefficients, free valence coefficients, and other suitable coefficients).

The structure of the receptor, on the other hand, is still poorly known. The procedure is to assign it arbitrary coefficients, which are indeterminate and take into account the known coefficients of the tetracyclines and experimental measurements of the antibiotic activity of five or six previously synthesized antibiotics; in this way, the coefficients of the receptor are determined by statistical methods. These coefficients can then be used to calculate *a priori* the activity of a tetracycline compound solely on the basis of its formula—on the assumption that its ability to penetrate the bacterium is similar to that of the five or six tetracycline compounds that were used as the basis for the calculation.

It will be obvious that this type of method can help direct the search for the most effective tetracyclines, and it has indeed yielded satisfactory results. This technique, which can be applied to the search for a great variety of drugs, is the foundation of quantum pharmacology.

THE ORIGIN OF CANCER CELLS

I began my research career in 1942 at the *Institut du radium*, working as an assistant to Professor Irène Joliot-Curie. One of the co-directors of the institute, the eminent cancer specialist Professor Antoine Lacassagne, had done important work paving the way for research on the role played by hormones in inducing breast cancers. Professor Joliot-Curie urged me to pay him a visit, since he had read an article by a German researcher, Otto Schmidt, who reported a relationship between the distribution of electron densities in conjugated hydrocarbons and their carcinogenic properties.

Remember that by this time, researchers already knew how to induce cancers experimentally using certain compounds. Let us describe a typical experiment. First we prepare a 0.3% solution of 9,10-dimethyl-1,2-benzanthracene in benzene. This compound is a conjugated hydrocarbon similar to the ones we discussed with reference to chemical reactivity. We then select a group of 40 mice and divide them into two subgroups of 20 each. The mice in one of these subgroups will act as controls, and will not be given any particular treatment. The other mice will have the conjugated hydrocarbon solution swabbed onto their backs twice a week. After about six weeks, a little more than half the mice in this second subgroup will have developed skin cancer on their backs. Therefore, 9,10-dimethyl-1,2-benzanthracene induces cancer; it is carcinogenic.

It should be obvious that a detailed understanding of the mechanism by which cancer cells form under the effect of various molecules is of the greatest importance in preventing and treating this disease. The presence of carcinogenic compounds has been demonstrated in smoked foods, charred food (the blackened parts of toast or grilled meats), and the tar formed by the combustion of tobacco. The risk of cancer can therefore be mitigated by not eating too many smoked or charred foods, and by not smoking.

In his 1941 article, Otto Schmidt reported the presence of an electron-rich region in certain carcinogenic conjugated hydrocarbons. I therefore directed the work of a young researcher in my group, Alberte Pullman, towards an investigation of these molecules. The quantum-mechanical calculations made by Otto Schmidt had not been very accurate. I repeated those calculations in 1945 in collaboration with Pullman, and in doing so we introduced the concept of the free valence coefficient. We then constructed, for a variety of compounds, molecular diagrams which graphically represented the bonding coefficients and free valence coefficients.

In 1946, using these diagrams, Pullman observed the presence of a pair of adjacent carbon atoms, highly likely to react by addition (producing what organic chemists call "addition reactions") in certain carcinogenic conjugated molecules. She gave the name "K region" to this pair of atoms. Now back in 1944, Antoine Lacassagne, Buu Hoi, Georges Rudali, and I had postulated, in order to interpret certain experiments, that carcinogenic conjugated molecules bind to certain important cell molecules such as proteins or nucleic acids. Pullman's observations confirmed the validity of this hypothesis, and for more than ten years starting in the 1950s, a number of researchers—especially the Americans E. C. Miller and C. Heidelberger, as well as Pascaline Daudel (my wife)—succeeded in isolating derivatives formed by addition reaction between a carcinogenic substance and certain proteins or nucleic acids extracted from animals treated with the substance.

Suddenly Heidelberger's team uncovered an important fact: they found that the carcinogenic conjugated hydrocarbons that bind with proteins and nucleic acids in animals (*in vivo*) do not attach themselves to those same molecules when they are isolated in the laboratory (*in vitro*). The obvious conclusion is that the conjugated hydrocarbons must be converted in some way by the living organism before reacting with its proteins and nucleic acids. I therefore suggested to my wife that she enlist every resource offered by the quantum molecular sciences in the investigation of this phenomenon.

At that time, two brilliant researchers at the *Institut du radium* had just developed a highly effective instrument for studying the nature of the fluorescent light emitted by molecules; this instrument could detect the energy of practically every photon that emerged from a molecule. Pascaline Daudel, M. Duquesne, and P. Vigny (with the assistance of two British researchers, P. Grovers and P. Sims) then demonstrated that it was not the conjugated hydrocarbons that attached to the proteins and nucleic acids of living cells. The organism, in an attempt to rid itself of these "exogenous" hydrocarbons, oxidizes them, and it is the oxidized compounds (diol-epoxides) that are the true carcinogens which bond to the proteins and nucleic acids. In its effort to eliminate these intruders, the organism commits a fatal error: it transforms them into formidable enemies.

We still needed to understand how this interaction with proteins and nucleic acids ends up inducing a cancer. First we must remember that a cancer cell has at least two peculiarities. In a normal adult organ, when a cell dies, another cell is born (except for certain cells such as neurons, which are almost never replaced). That is why the weight of an organ in a healthy human being remains approximately constant throughout its owner's adult lifetime. This fact shows that a cell divides in two only if it receives an order to do so. But cancer cells become insensitive to these orders and begin dividing anarchically: that is the origin of the primary tumor.

The second peculiarity: from the embryonic stage onward, our cells differentiate (although they all contain the same genetic message). A brain neuron is very different in shape and molecular inventory from a muscle cell. What happens is that in each cell the genetic message is implemented in only one specific way: certain genes work very hard while others, previously active in the embryo, become inactive in the adult. In other words, each type of cell contains a genetic program that implements one part (and only that part) of the genetic message. This differentiation ensures that we have no liver cells in our brain!

But cancer cells lose some of that differentiation: a prostate cancer cell can migrate into a bone and give rise to a tumor affecting that bone. In technical terms, the primary tumor is said to produce secondary tumors (or "metastases") in other organs. Several years ago, researchers began to understand how carcinogenic compounds destroy cell differentiation. Genetic activity is controlled by small segments of DNA called operators. By attaching to the DNA, carcinogenic molecules can modify these operators, making them more active and awakening dormant genes which create the differentiation among cells.

THE AIDS VIRUS

The development of treatments for cancer meets the definition of what international organizations like UNESCO refer to as a "world problem." The same is true today, of course, for the fight against AIDS. And once again the quantum molecular sciences are making a contribution which may be decisive in overcoming this scourge.

Like all living organisms, human beings defend themselves against intruders with a powerful immune defense system. If a virus, bacterium, fungus, protozoan, helminths, or cancer cell attacks us, an army of our cells goes into action and declares war on it. The great majority of these cells consists of white cells in the blood and lymph. Assume that a virus penetrates the barriers that protect us (the skin, for example), and enters the bloodstream. A white cell called a "macrophage" will destroy the virus by phagocytosis, that is by absorbing it and breaking it into pieces, specifically into molecules called antigens.

In a second phase, the macrophage will present one of these antigens to another type of white cell called a T4 lymphocyte, which is capable of recognizing it. This recognition is based on intermolecular forces between the antigen and a receptor on the lymphocyte. When recognition has occurred, the macrophage sends the lymphocyte certain messenger molecules called "interleukin 1," which stimulate the T4 lymphocyte that has been activated by the antigen. The T4 lymphocyte in turn activates two other types of lymphocytes, prompting them to multiply by sending them a whole series of messages in the form of various molecules, especially "interleukin 2." The first type of lymphocyte is called T8: these are killer cells, which attack cells in our body that have become infected by the virus, and allow the macrophages to destroy them. The second type of lymphocyte—the B lymphocyte—produces molecules called antibodies, which attach themselves to the virus and neutralize it. The body also forms memory cells which, once they have recognized the virus, are well-prepared to fight it the next time it appears inside the organism. This phenomenon is the basis for the effectiveness of vaccines.

But something entirely new has now happened. About twenty-five years ago in Africa, and a little more than a decade ago in the United States and Europe, there appeared a virus capable of destroying our immune defenses: it is called the AIDS (acquired immune deficiency syndrome) virus. This virus—or rather these viruses, since there are two strains known as HIV1 and HIV2 (for human immunodeficiency virus)—attaches to a specific membrane receptor called CD4, which is present on T4 lymphocytes and macrophages. In a process that is now well understood, the viruses use this receptor to penetrate into cells on which it is present. The viral RNA then uses a molecule called reverse transcriptase to produce a copy of its genetic message in the form of DNA. This DNA fragment, the "provirus," combines with the cell's own DNA and causes the cell to produce copies of it, which develop into new viruses. The HIV virus thus multiplies primarily by infecting T4 lymphocytes and macrophages, gradually causing these cells to die out. Little by little, the virus destroys the cellular army that constitutes our entire immune defense. Deprived of its defenders, the body falls prey to all sorts of conventional parasites: helminths, protozoans, other viruses, bacteria, fungi, cancer cells, etc. Many so-called "opportunistic" diseases thus appear in people infected with the AIDS virus. Often very difficult to treat, some of them cause terrible suffering and death. In AIDS patients, certain body fluids—primarily blood, semen, and vaginal secretions—contain large numbers of "virions" (mature stages of the virus). The primary mechanisms of transmission are therefore introduction into a healthy organism of a small quantity of infected blood, and sexual relations involving exchange of these body fluids (between persons of the same or opposite sex).

The problem of AIDS is the subject of international cooperation which requires better coordination. For example, the European Academy of Sciences, Letters, and Arts, established in 1980 thanks to the initiative of Nicole Lemaire d'Agaggio, has joined this fight as part of its commitment to world problems. In association with UNESCO, the

Academy—backed by its 200 titular members, selected from some fifty national academies, and its 200 corresponding members, and with the support of the European AIDS Research Foundation—has organized a group of nine "Euro-American" research laboratories. This group, known as "Humanity Against the Virus" and chaired by Professor Luc Montagnier, is exploring three key research directions in the fight against AIDS: development of a vaccine; improved treatment; and analysis of AIDS pathogenesis.

An example of the research being performed within this group will show once again the importance of gaining an accurate understanding of molecular behavior. Antiviral vaccines are often prepared from attenuated or inactivated virus. But the AIDS virus evolves very rapidly, and can remain latent for a very long time. There is reason to worry that an inactive strain might easily become reactivated. The Humanity Against the Virus group therefore decided to use the "immunosome" technique developed in one of the group's laboratories led by Professor Lise Thibodeau in Canada. The method involves replacing the protein coat that surrounds the core of the virus (known as the "capsid") with droplets of a phospholipid (a synthetic molecule) and hooking molecules of antigens extracted from an HIV virus onto the surface of these spheres (called liposomes). The resulting immunosomes cannot infect a human being because they contain no DNA and therefore no genetic messages.

However, provided matters are arranged so that the antigen molecules on the surface of the liposome retain the same form that they have in the virus, these immunosomes are theoretically capable of triggering the body's immune defenses just as the virus would. Experiments on mice have already confirmed that this vaccine induces an immune response, and attempts are now being made, using monkeys, to measure the protective effect of this potential vaccine. We can now see clearly that once it has been recognized, explored, and mastered, the realm of molecules has nothing tyrannical about it: on the contrary, we should look upon it in a spirit of freedom and hope.

CONCLUSION:
THE DANCE OF
THE MOLECULES

Let us take a moment now to look back on the major facts we have discussed. We know that the properties of the inanimate objects that surround us—gases, solids, matter of all kinds—depend on their constituent molecules or atoms and the ways in which they are arranged. Living organisms are the result of a highly complex molecular organization: the fact that they are alive is primarily the consequence of a multitude of chemical reactions. If one of those reactions goes awry, the organism is in danger. Much of our future is foreordained in the fertilized egg with which we begin at the moment of conception. If, at that instant, our twenty-third pair of chromosomes is XX, we will be female; if it is XY, we will become male. If our twenty-first set of chromosomes is triple, we will be born with Down syndrome. The color of our skin and texture of our hair, our height and build, and much more, depend largely on our genome, in other words an assemblage of large molecules. We also know that, unfortunately, some of the diseases that we risk contracting also have to do with that same genome.

The operation of our brain depends on an infinity of entangled tiny electrical currents, controlled by multiple chemical reactions. Our eyes function because the light photons received by our retinas change the shape of the retinal molecules contained in them. Our muscles would not contract without electrical currents that begin in the brain and travel along nerves to the neuromuscular endings, where a neurotransmitter molecule (acetylcholine) changes the length of our muscle fibers. Our emotional life and our most intimate feelings are accompanied by a host of chemical phenomena. To a large extent, molecules and molecular laws therefore account for everything that exists and lives on our planet.

CAN WE CONTROL THESE MOLECULES?

If we want to exert some kind of mastery over the world around us, we must learn to control molecules. If we understand the molecular world, we gain the ability to manufacture an endless series of new materials, and to create more and more powerful machines, such as the computers which are revolutionizing the world of information processing. By removing a few cells from a human embryo while it is still in the womb, we can determine whether it will become a girl or a boy, and whether it runs the risk of being born with a hereditary disease. Thanks to our understanding of the molecular world, we can synthesize drugs to help cure numerous ailments.

As we have seen, the quantum molecular sciences allow us to set up a mathematical government within molecular populations; and we have learned that these same quantum sciences impart a very specific structure to our knowledge. The first principle of quantum mechanics states that every variable (the mass, energy, velocity, etc. of a particle) is associated with an operator. Every possible magnitude of the variable is one of the eigenvalues of the operator. To draw up a list of the possible magnitudes of a variable, for example the energies of a molecule's quantum states, we therefore have the choice between making measurements in the laboratory and performing calculations with a calculator (or, more efficiently, with a computer).

The first principle creates a correlation between laboratory operations (which are both theoretical and experimental, since experimental results cannot be interpreted without theory) and purely mathematical operations. Solving a given problem usually involves taking into consideration a set of variables, which will correspond to a set of operators.

When we look more closely, we find that the sum of two variables is associated with the sum of the operators, and that the product of two variables corresponds to the product of two operators. The set of variables comprises two operations on those variables, and that constitutes an algebraic structure. The set of operators comprises analogous operations, and thus constitutes another algebraic structure. From a mathematical standpoint, the first principle establishes what we have come to call an "isomorphism" between a set of variables (leading to laboratory operations) and a set of operators (leading to mathematical operations).

The second principle of quantum mechanics, which is embodied in the wave equation, can be used (by means of mathematical calculations) to predict changes in the wave associated with a system of particles as it is studied over time under certain specific experimental conditions.

The third principle allows us, on the basis of this wave, to calculate the probability that the magnitude of one of the system's variables is equal to one of the eigenvalues of the associated operator. It then becomes possible, for example, to calculate the probability of finding that a hydrogen atom, which was in its ground state before a photon struck it, is now in its first excited state. In other words, the third principle permits us to calculate the probability that the photon will be absorbed by the atom. And we have seen that such calculations offer a way of understanding or predicting the colors of atoms and molecules, and thus of the objects around us.

In sum, the three principles of quantum mechanics provide a highly precise structure for scientific knowledge, based on an isomorphism between two sets; a wave equation; and a probabilistic result. The objects that interest us contain multitudes of molecules, and we know that according to the law of large numbers, multitudes of probabilities will combine at this level to create certainties.

The volume of calculations that must be made to solve a problem on the basis of quantum mechanical principles can be intimidating. Fortunately, the advent of computers has given us the means to perform a large number of mathematical operations in a short period of time.

In 1960 I was in Chile along with Roland Lefebvre, a member of my laboratory. We wanted to train a group of eight Chilean students to perform quantum mechanical calculations. The assignment was to calculate certain variables associated with the carbon monoxide (CO) molecule, in particular the energy of its ground state and the mean distance between its nuclei. At that time the University of Chile did not own a computer, and all we had available to us were a dozen small desktop adding machines. To achieve our goal, the ten of us—eight students and two teachers—had to spend an entire month doing several million additions, subtractions, and multiplications!

And we always had to work in pairs, duplicating the same calculations as a check on our work; even with practice, a mistake is made about once every hundred operations. When we returned to my laboratory in Paris, Lefebvre and I repeated the calculations on a computer. This undertaking, which had required a month of hard work by ten people, took only five minutes this time; and all false modesty aside, we and our students had not made a single mistake.

THE QUANTUM SCIENCES AND THE ARTS

Most painters and sculptors of past centuries have drawn inspiration for their work from scenes of daily life, historical events, or religious writings. Only a few iconoclastic artists limited themselves to the use of geometrical structures. But at the beginning of the 20th century some entirely new artistic initiatives arose. Relationships between the sciences and the arts multiplied and diversified, and a true metamorphosis took place in art. Painters such as Robert Delaunay, Vassily Kandinsky, Kasimir Malevitch, and Piet Mondrian threw themselves into the audacious international adventure of abstraction. After World War II this abstract art (or rather this non-figurative art) slowly began to coalesce around two main orientations.

The first, called constructivist abstraction, involved the creation of paintings from simple geometrical figures: squares, triangles, diamonds, and rectangles. In the long term, this approach dissolved into geometrical irrationality. The second, more gestural movement (non-constructivist abstract art), covers a variety of diverse schools, in particular *abstraction lyrique* as represented by Hans Hartung, Georges Mathieu, and Wolfgang Wols, as well as Jackson Pollock and the "action painters," of whom I consider Henri Michaux a good representative.

In the meantime, around 1910, two important artistic currents had appeared. These were Rayonnism, led by Mikhail Fyodorovich Larionov, which consisted in embodying on canvas the luminous energy emitted by objects; and Futurism, represented especially by the Italian artists Giacomo Balla, Umberto Boccioni, Carlo Carrà, Rossolo, and Gino Severini, whose essential preoccupation was to bring matter to life by translating matter into movement.

Towards 1920, Naum Gabo created his first kinetic sculpture: a vibrating blade driven by a motor, which defines virtual volumes in space. Then Alexander Calder turned to the wind to move his mobiles; Jean Tinguely's metamechanical constructions clanked to life; Chryssa began to sculpt with luminescent tubes; Cosmas Xenakis organized his impressive polytopes around traceries of laser beams; and Nicolas Schöffer constructed his remarkable cybernetic sculptures. And now computer-assisted art—"infography"—is taking flight. This is the age of the synthesizer: Roger Lafosse's "cortica-lart" allows composers to improvise by direct use of the encephalographic waves generated in their brains. René Huyghe has said that artists are "reveal-ers of the collective unconscious" of our society.

At a time when our daily life is being more and more radically transformed by technologies created by science, it is normal—in fact it is most encourag-ing—to see how many works of art are using scientific techniques. It was only to be expected, then, that the quantum sciences would make an appearance in the world of art.

To conclude, I would like to mention the works of two artists who have indeed drawn inspiration from this field that is thought to be so highly abstract. One of them, Bettina Brendel, is an American painter who met several times with the famous German quantum physicist Werner Heisenberg. Brendel has produced several large works created by juxtaposing multiple paintings on the theme of "particles or waves." Her purpose, in other words, has been to evoke the duality—the point-like and wave-like nature—of elementary particles like the electron. The artist has constructed her pictures from geomet-rical elements that suggest these two aspects of the infinitesimally small. So we see in them a multitude of tiny lines, perhaps recalling brief trajectories or the rays associated with particles. We also find sequences of waves and shapes evoking what a particle "sees" as it passes through our optical or electron microscopes. Brendel's work thus has roots in constructivist abstraction, but it is based on a theme of scientific origin.

Nicole d'Agaggio has gone further. She has created an enormous painting, over 7 feet high by 23 feet long, inspired by the three great principles of quantum mechanics that we discussed earlier. To paint this canvas, the artist drew on the extraordinary flexibility of the symbols of lyrical abstraction. In the background of the painting appears a subtle rank of evanescent columns, like a temple erected to the glory of quantum mechanics. The wave equation asserts itself at the center of the picture, like Japanese calligraphy. On the left, the artist recalls the first principle; we see the "variables" and their shadows which symbolize the "associated operators." On the right we find a trace of the third principle: a wave vector projected onto three diffuse axes.

Traversing the entire length of this extraordinary work is a beam of light that diffracts as it falls on a screen pierced by two holes: Young's famous experiment. Copper, gold, and bronze, like burning embers, shoot furtive gleams into space, creating a symphony of light whose appearance changes as the spectator moves. All these optical shifts remind the viewer that quantum mechanics is the culmination of much thought about the profound nature of light.

This remarkable painting gives a very good idea of that "great rational beauty" that Louis de Broglie had already sensed as he contemplated the nascent structures of wave mechanics.

In 1973, I was given the responsibility of chairing the First International Conference on Quantum Chemistry, held at the Palais de l'Europe in Menton. When I learned that one of the municipal councillors of that town was a dancer at the Stuttgart Opera, I had the idea of presenting, at the Conference's opening session, a brief choreographic entertainment retracing the history of the concept of the electron. The music that I used for the basic elements of the choreography was an organ improvisation by the fine French composer Jean Guillou.

The classical electron, as it was envisioned before the coming of quantum mechanics, is evoked in the first part of the ballet. The tutu-clad dancer moves to the steps of romantic choreography, rotating rapidly on her own axis to suggest the spin of the electron. The second part of the dance deals with the electron in quantum mechanics, where its trajectory can no longer be followed. The dancer has discarded her tutu, and leaps suddenly to and fro in flashes of light, her movements inspired by the choreography of Maurice Béjart. Spotlights cast a shadow of the dancer on the backdrop to symbolize the associated operator, which is also evoked musically by the pedal rank of the organ. The final interlude is an apotheosis representing the eigenvalue equation of the spin operator, in which the dancer uses her body to write the Greek letter Ψ, the symbol for the wave associated with the electron.

BIBLIOGRAPHY

BUSCH, HARRIS, *The Molecular Biology of Cancer*, Academic Press, 1974.

CHANGEUX, JEAN-PIERRE, *L'Homme neuronal* [The human neuron],* Fayard, 1983, new ed. "Pluriel," 1987.

CHRISTEN, YVES, *L'Homme bioculturel* [Human bioculture],* Éditions du Rocher, 1986.

DAUDEL, PASCALINE and DAUDEL, RAYMOND, *Chemical Carcinogenesis and Molecular Biology*, Wiley, 1966.

DAUDEL, RAYMOND, *Vision moléculaire du monde* [A molecular view of the world] (with reproductions of works by the painter Nicole d'Agaggio),* *Centre national de la recherche scientifique* (CNRS)/Hachette, 2nd ed., 1990.

FRAGA, SERAFIN, *Quimica Teorica* [Theoretical chemistry],* CSIC, 1987.

GROS, FRANÇOIS, *Les Secrets du gène* [The secrets of the gene],* Odile Jacob, 1986.

GUINIER, ANDRÉ, *La structure de la matière* [The structure of matter],* CNRS/Hachette, 1980.

KORNBERG, ARTHUR, *DNA Synthesis*, Freeman, 1974.

RICHARDS, WILLIAM GRAHAM, *Quantum Pharmacology*, Butterworths, 1977.

WEISSKOPF, VICTOR, *La Révolution des quanta* [The quantum revolution]* (*Questions de science* series), Hachette, Paris, 1989.

---

* These references have not been published in English.